Calculus

D1471635

Studymates
Helping You to Achieve

Calculus
How Calculus Works

Christine Tootill

© 2008 by Christine Tootill
additional material © 2008 Studymates Limited.

ISBN: 978-1-84285-079-4

First published in 2008 by Studymates Limited.
PO Box 225, Abergele, LL18 9AY, United Kingdom.

Website: http://www.studymates.co.uk

Typeset by Vikatan Publishing Solutions, Chennai, India
Printed and bound in Europe

Contents

Advice to Students

Are you studying calculus? Then this book is certainly for you. By breaking the topic down into easily managed chunks, this book provides real help for students of mathematics aged 16 to adult who are working towards university entrance or following any other course at a similar level.

Calculus is often thought to be more difficult than other topics in mathematics. This may be because it is completely new to you, the student, at this level. We have therefore included plenty of examples for you to use as practice – which of course is the best way to overcome these apparent difficulties.

This book takes you through a step-by-step approach and thoroughly introduces you to the topic of calculus. It also provides opportunities to revise the foundation topics you need for calculus – the Appendix is a mini-revision book and there are also some recommended websites for you to use.

We have focused on the basics and taken as practical an approach as is possible in pure mathematics so this book will enable you to understand the fundamental concepts and achieve success with examination questions.

Mathematics is a conceptual subject. It takes time to absorb new concepts, so don't try to rush through this book. It would be a very good idea to work through each chapter twice before you move on. Sometimes you may find you get stuck on a particular question, or you may not understand a particular point on first reading, but this doesn't mean you have to stop at that point.

Work through the rest of the chapter and then go back to whatever caused you a problem. If you still find it difficult, leave it, and move on. It is not always necessary to understand every word of a book in order to gain an overall understanding of the topic in hand. Also, it is not necessary to have answered every single example correctly before progressing to the next chapter. You will probably find that if you return to points of difficulty at a later date, you will by then have absorbed the ideas better and be able to cope with what previously seemed difficult. You are the only person who knows whether you have understood a topic and are ready to move on – so trust your instincts on this.

Do work with other students if you have the opportunity. It is an established fact that students can learn as much from each other as they can from teachers and books. Mutual support also dispels the notion, common to many students, that they are the only person who is finding the work hard and everyone else is cleverer than they are!

Christine Tootill

1 The meaning of change

One-minute summary

This chapter is a very gentle introduction to one of the basic mathematical concepts which you need to understand before you start work on Calculus. Here we look at relationships between changing quantities and introduce notation which is essential in Calculus. This chapter assumes you are familiar with equations of straight lines and the calculation of gradients. If you need to revise these topics, do this first. Use the Appendix (pages 81–106) in this book or refer to the end of the websites list (Page 139).

Working with changing quantities

Anyone who has studied mathematics has at some time come across the phrase "rates of change".

A rate of change is simply a measure of how much a quantity increases (or decreases) by, in relation to another quantity.

Example 1.1

Suppose apples cost 80 cents per kilo. Then 2 kilos of apples cost $1.60, 3 kilos cost $2.40, and so on. We can draw a graph to represent this information:

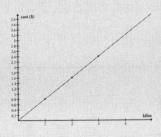

Fig 1.1

In this example the quantities which can change are the amount you buy (the number of kilos) and, as a result, the amount of money you spend. The rate of change here is the price per kilo, 80 cents. The amount of money you spend increases by 80 cents for each extra kilo of apples you buy.

The gradient of the line in Fig 1.1 is $0.8 = 80$ cents. In a linear relationship, the gradient of the line always corresponds to the rate of change.

Example 1.2

The population of a small town in the year 2000 is 4000 people. Starting in 2000, 100 new houses are built in the town each year. Suppose that the average number of occupants of each new house is 4. Then the new houses are adding 400 to the population of the town each year.

The rate of change in this situation is the increase in the population per year, which is 400.

The actual rate of increase of population in the town will be less than this, because some people die and some move away. Suppose the decrease in population from these two causes is 100 per year; this is another rate of change, and since it is a decrease, we write this as -100 per year.

What then is the actual rate of change of population in this town? If we combine (i.e. add) the rate of increase and the rate of decrease, then the net effect is that the town is gaining $400-100 = 300$ inhabitants per year.

This example is important because it illustrates how different rates of increase or decrease relating to the same situation can be combined to find the overall rate of change. Population studies are in fact one of the areas of mathematics to which calculus can be usefully applied.

Practice questions (1.1)

[Note – Answers to all questions in the text are given in the Answers section which starts on p. 107]

1. Using D to represent dollars and k to represent kilos, write down the equation of the line in Figure 1.1.

2. This question relates to Example 1.2 above.
  (i) Draw a straight line graph showing the population increase due to house-building.
      Remember that the population starts at 4000 in the year 2000. Mark the years as 0, 1, 2, 3 on the horizontal axis, with 0 representing the year 2000.
  (ii) On the same graph, draw a straight line showing the population decrease due to people dying or moving away.
  (iii) On the same graph, draw a straight line showing the actual population change.
  (iv) Write down an equation for each of these 3 straight lines. Let P represent the population and Y represent the year, so $Y = 0$ is the year 2000.
  (v) Check that in each case, the gradient of each line is equal to the rate of change (i.e. increase or decrease) of the population.

Notation

In calculus, we use a shorthand notation for the gradient of a line, or the rate of change.

The gradient of a line is calculated by taking any section of the line, finding the differences between the vertical and horizontal co-ordinate values, and dividing the vertical difference by the horizontal difference. On an (x,y) graph, the gradient will be:

$$\frac{\text{difference in } y}{\text{difference in } x}$$

We abbreviate this to $\frac{dy}{dx}$ where the 'd' stands for difference.

The variables will be represented by various letters in different problems, but the 'd' is always used in this notation, so it is sensible to avoid using lower case 'd' as a variable.

Practice questions (1.2)

1. Use the gradient notation to write down the gradients of each of the three straight lines in Example 1.2

2. On a graph with time t on the horizontal axis and distance s on the vertical axis, what will the notation for the gradient be? What will this rate of change represent?

3. The cost C of hiring a car in Italy is €50 (fixed charge) plus €20 per day.
   (i) Let n be the number of days' hire, and write down an equation connecting C and n.
   (ii) What is $\frac{dC}{d\epsilon}$ and what does it mean?

4. Water is dripping at a steady rate onto a horizontal surface and creating a circular puddle.
   (i) If the radius of the puddle is r, write down the formula for C, the circumference of the puddle.
   (ii) What is $\frac{dC}{dr}$ and what does it mean?

5. The time taken to cook a chicken is given by the formula $T = F + 30k$ where k is the weight of the chicken in kilos and F is a fixed number of minutes.
   (i) Write down the value of $\frac{dT}{dk}$ and explain what this means.
   (ii) If it takes 1¼ hours to cook a chicken which weighs 2 kilos, find the value of F.

6. The general equation of a straight line is $y = mx + c$. If you are given a numerical value for dy/dx, can you:
   (i) find the value of m?
   (ii) find the rate of change of y with respect to x?
   (iii) find the value of c?

Tutorial

Progress questions (1)

1. A cookery book gives the following table of cooking times for a turkey:

W (weight in kilos)	1	2	3	4
Time (minutes)	45	65	85	105

   (i) Find a formula for T in terms of W.
   (ii) Find how long it will take to cook a turkey weighing 6 kilos.
   (iii) Find the value of $\frac{dT}{dW}$ and explain what this means.
2. The brakes of a car, when fully applied, cause it to lose speed (i.e. decelerate) at a rate of 15 mph per second.
   (a) How many seconds will it take for the car to come to a standstill if the brakes are applied when the car is travelling at (i) 60 mph (ii) 42 mph?
   (b) Let V represent the speed of the car. Let t represent time in seconds. At the moment when the brakes are applied, the speed of the car is u, and $t = 0$.
      (i) Write down an equation connecting V and t.
      (ii) Find dV/dt and explain what it means.

[Note: In a real situation the relationship between speed and stopping time is not linear, especially at higher speeds. Do not assume the numbers in this example are correct when driving!]

Practical assignment

Try one of these experiments.

You will need either a cooking thermometer or an oven with a temperature display.

Experiment 1

Boil some water and pour about half a pint into a suitable container. Set the time, t, to zero.

Measure the temperature of the water every five minutes and write your results in a table:

t	0	5	10	15	20
T	100				

Plot your results on a graph.

Is the relationship between T and t a linear relationship, i.e., is it possible to find values of a and b so that $T = a + bt$?

If the answer to this question is yes, then write down the rate of change of temperature with respect to time. This is equal to dT/dt.

If the relationship is not linear, then the graph of this data will not be a straight line, and the gradient will not have a constant value.

Experiment 2

Set the oven to heat to its maximum temperature. Record the actual temperature every 5 minutes and write your results in a table. Continue as in Experiment 1.

Seminar discussion

"Rates of change in real life situations are more commonly positive (increasing) than negative (decreasing)." Suggest reasons why this statement may or may not be true.

Study tip

The main objective of this chapter is to achieve an understanding of the concept of **rate of change**.
If you are not completely clear about this, revisit the examples at the beginning of this chapter and repeat at least two of the Practice Questions.

2 Exploring curves and gradients

One-minute summary

A straight line has a gradient which is the same at any point on the line. By definition, a curve changes gradient at every point. Gradients at different points on a curve can be observed visually (positive, negative, small, large). Their numerical value can be estimated by drawing, or they can be calculated accurately. In this chapter we will
- revisit curves which are familiar to you
- examine their gradients

What will calculus do for us?

The techniques of calculus enable us to extend our mathematical skills beyond straight lines to curves. For example, when we are dealing with travel at a constant speed, the graph of the journey will be a straight line, and we can find the speed by the formula:

$$\text{speed} = \text{distance/time}.$$

If you want to revise the topic "Speed, Distance and Time" see section 2 of Chapter 14.

Differentiation (the first technique of calculus) enables us to find the *instantaneous* speed (i.e., the speed **at any moment** on a journey), provided that the journey can be represented by a curve whose equation we know.

Similarly, we can work out the area of a triangle using simple arithmetic, but if we want to find the area of a shape with a curved edge, we need to use *integration* (the second technique of calculus). Again, we would need to know the equation of the curve at the edge of the area.

In this section we look at three familiar curves:

- The quadratic curve $y = x^2$
- The cubic curve $y = x^3$
- The rectangular hyperbola $y = 1/x$

The quadratic curve

The graph of a **quadratic function** is a *parabola* – this is the name given to the shape of the graph. The simplest quadratic function is $y = x^2$ and its graph looks like this:

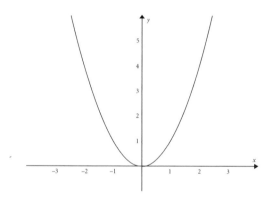

Fig 2.1

The gradient at any point on a curve will not, in general, be the same as at any other point.

Practice questions (2.1)

Look at the shape of the quadratic curve and answer these questions:

1. What is the gradient of the curve at the origin?
2. In which section of the graph is the gradient negative?
3. Where on the graph is the gradient large, i.e. steep?
4. Are there two or more points on this curve which have the same gradient?

The cubic curve

The graph of the simplest **cubic function**, $y = x^3$, looks like this:

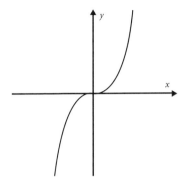

Fig 2.2

The rectangular hyperbola

The equation of the simplest **rectangular hyperbola** is $y = 1/x$, and its graph looks like this:

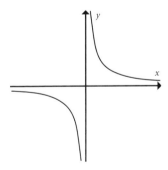

Fig 2.3

Practice questions (2.2)

Study the shapes of the cubic curve and the rectangular hyperbola, and answer Practice Questions (1.1) for each curve.

Tutorial

Progress questions (2)

1. Sketch a curve with the following properties:
   - the curve passes through the origin
   - its gradient is zero at the origin
   - the curve occupies the 3rd and 4th quadrants only
   - its gradient is positive in the 3rd quadrant and decreases as x increases
   - its gradient is negative in the 4th quadrant and decreases as x increases

Suggest a possible equation for your curve.

[Notes: When a gradient is **negative and decreasing**, the value of the gradient becomes **more negative**, (further away from zero), and the curve gets steeper. When a gradient is **positive and decreasing**, the value of the gradient is getting nearer to zero, so the curve gets less steep.]

2. Sketch a curve with the following properties:
   - the curve passes through the origin
   - its gradient is zero at the origin
   - the curve occupies the 2nd and 4th quadrants only
   - its gradient is negative in the 2nd quadrant and increases as x increases
   - its gradient is negative in the 4th quadrant and decreases as x increases

Suggest a possible equation for your curve.

Practical assignment (2)

Sketch the curves of the functions $y = x$, $y = 1/x$ and $y = x + 1/x$.

Look at the gradient of each function at the point where $x = 1$.

Choose another point and examine the gradients.

What do you deduce about the relationship between the gradients?

Seminar discussion

Is it possible to define the gradient at every point on every curve?

Study tip

Practise sketching the curves we have looked at in this chapter. You should be able to do this without reference to books. Learn what you have discovered about the gradients of these curves.

3 How to calculate gradients

One-minute summary

The essence of differentiation is best seen by working through a "frame-by-frame" process which will remove the mystery from this topic, but which you will later do in one step. This chapter shows you that process and provides you with some exercises to reinforce your understanding. These may seem laborious, but **be patient** – you will be rewarded with a clarity of understanding which will support you through your continued study of calculus. In this chapter we work through these stages:

- the gradient of a straight line
- the gradient of a tangent
- the gradient of a chord
- the gradient of a curve

Rates of change on straight line graphs

A straight line has a constant gradient. This gradient is the *rate of change* of y with respect to x. For example, if we plot a distance-time graph for a journey in which the speed stays constant, and we use the x-axis for time and the y-axis for distance, then the graph will be a straight line. The gradient of this straight line graph will represent the speed, which is the rate of change of distance with respect to time.

The mathematical relationship between speed, distance and time is explored in Section 1 of Chapter 14 (revision topics).

We can calculate the gradient of a straight line by placing a right-angled triangle at any position on the straight line and dividing its height by its width. This is illustrated in Fig 3.1:

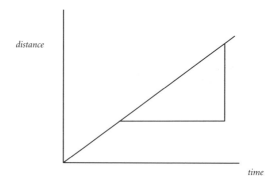

distance

time

Fig 3.1

You will probably be familiar with this from your elementary mathematics, but if you want to revise the topic "Straight lines and gradients" see section 2 of Chapter 14.

Gradient of a tangent to a curve

Calculus enables us to extend the above process to cases where the rate of change varies.

Suppose a journey starts slowly and the speed gradually increases. The distance-time graph for such a journey will look something like this:

We can *estimate* the speed at one particular point on the curve, by drawing a **tangent** at that point on the curve

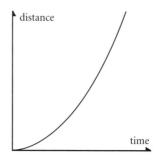

distance

time

Fig 3.2

and calculating the gradient of the tangent. A **tangent** is a straight line which **touches** a curve at a particular point and has the same gradient as the curve at the point of contact.

But this method will only give an *estimate* because in drawing a tangent free-hand we cannot ensure that the gradient of the tangent is exactly the same as the gradient of the curve at the point where the tangent touches it.

From the gradient of a chord to the gradient of a curve

We can find the *average* speed between two points on this curve, by joining the two points with a chord and calculating the gradient as for a straight line graph. The method of calculus is based on this idea, but takes it a crucial stage further by shrinking the chord until its two ends meet. The chord is thus replaced by a tangent. This process is called "taking the limit" and you can do this by working through the next set of questions.

Practice questions (3)

a) Using a large sheet of graph paper and a large scale on both axes, plot the curve $y = x^2$ for positive values of x and y only. Label the points A (1,1) and B (2,4). Draw the chord AB and find its gradient.

Now choose a new position for the point B between A and the first B. Draw the chord AB and find its gradient.

b) Repeat at least three more times, with each new position of point B closer to A. Make a table of values showing for each point B, the co-ordinates of A and B and the gradient of AB.

c) Write down your guess for the gradient of the curve $y = x^2$ at the point (1,1).

d) Sketch the curve $y = x^2$ in the second quadrant ($x < 0$, $y > 0$). Write down your guess for the gradient of the curve at the point $(-1,1)$.

It's the limit!

You have now "taken the limit" of the gradient as the point B approaches the point A, i.e. as the chord AB approaches the tangent at A. You have taken the final "leap" by making an intelligent guess. You can find the complete mathematical procedure for this (known as *differentiation from first principles*) at:

http://www.mathsrevision.net/alevel/pages.php?page=23

but we will not include it here.

In the Answers section, on page 111, you will find a table of values showing the gradient of AB for various different positions for B. This is the work required in Practice Questions (3.1), extended to include points too close to A to be plotted on your graph.

You will see from the table that the gradient of AB approaches 2 as B approaches A. This is the limiting value – the gradient of the curve at point A. If you add more entries to the table, with B closer to A, you will see the gradient get closer to 2. But of course you cannot choose a position for B which gives the gradient exactly equal to 2. If you reduce the chord to a point by setting $B = (1,1)$ then you cannot calculate the gradient because you will be trying to divide by zero. The process of differentiation from first principles deals with this problem – see the website referred to above.

Tutorial

Progress questions (3)

1. (a) Plot the curve $y = x^2$ on a fresh sheet of graph paper (or use your graph from Practice Question (2)). Label the points $A = (2,4)$ and $B = (3,9)$. Draw the chord AB and find its gradient. As before, bring B closer to A and calculate the gradient of AB for each new position of B. Then write down the gradient of the curve at A.

   (b) Repeat (a) with $A = (-3,9)$ and $B = (-4,16)$. Write down the gradient of the curve at A.

2. Complete this table:

x	gradient (dy/dx)
1	2
2	
−3	

3. Deduce, from this table, the relationship between x and dy/dx for the curve $y = x^2$.

Practical assignment (3)

You will need a graphical calculator for this work. (Alternatively, you can solve the problem by algebra). Use your calculator to display the curve $y = x^2$ and the straight line $y = 3x - 2.25$ in the range $x = 0$ to $x = 5$. Use the trace facility on your calculator to find the co-ordinates of the point where the line touches the curve. Do your results confirm the relationship between x and dy/dx?

(If you want to revise the topic "Finding points of intersection" see section 3 of Chapter 14.)

Seminar discussion

Some mathematicians consider the notation dy/dx to be clumsy and misleading. But it has the advantage of reminding us that a gradient is found by division. Should we use a simpler notation for the gradient?

Study tip

The process in Practice Questions (3) is crucial to your understanding of differentiation. Revise what you have learnt so far by repeating this process at different points on the curve $y = x^2$.

Confirm that your results agree with what you found in Progress Questions (3) in this chapter.

4 How gradient functions work

One-minute summary

In this chapter we see that
- The gradient of a curve can be expressed in general terms, as a function of x
- Once we have found this function, we can find the gradient at any point on the curve
- We can then build up a set of results which will lead (in Chapter 5) to a general formula for the gradient function

The gradient function of the quadratic curve

We have seen that for the quadratic curve $y = x^2$, the gradient $dy/dx = 2x$.

This expression for the gradient is itself a function – it takes different values according to the value of x. We call this a **gradient function**. In this case it is a straight line with gradient 2, passing through the origin.

This is our first important result:

> The gradient function of the quadratic curve is a straight line.

Now we will investigate the gradient function of the cubic curve $y = x^3$.

Practice questions (4.1)

(a) Find the gradient of the curve $y = x^3$ at the point $A = (1,1)$. Use the same process as before, but start with

B close to A. It's not essential to plot the graph this time; you can just draw up a table of values for the co-ordinates and the gradient. Then write down the gradient of the curve at A.

(b) Repeat (a) with $A = (2,8)$.

(c) Repeat (a) with $A = (-3,-27)$.

Check your answers to these exercises before you move on.

We could continue to use this method for finding the gradient at any point on a curve, but this would be very slow and laborious. We need a **general** method, i.e. a formula which we can use to find the gradient without having to calculate the gradients of the sequence of chords.

We have seen that for the curve $y = x^2$, $dy/dx = 2x$. We can use this gradient function to find the gradient at **any** point on the curve $y = x^2$ by substituting the value of the x-coordinate into $dy/dx = 2x$. Look back at Practice Question (1). We observed from the quadratic curve that the gradient is negative when x is negative, zero at the origin, and large when x is large. Our gradient function $dy/dx = 2x$ confirms these observations.

Practice questions (4.2)

1. Compare the curve $y = 2x^2$ with the curve $y = x^2$. How does doubling the y co-ordinates affect the gradient? Deduce the gradient function of the curve $y = 2x^2$.

2. a) What is the gradient function of the line $y = 3x$?
   b) What is the gradient function of the line $y = -x$?
   c) What is the gradient function of the line $y = kx$?
       (k is a constant)

3. a) What is the gradient function of the line $y = 2$?
   b) What is the gradient function of the line $y = -1$?
   c) What is the gradient function of the line $y = a$?
       (a is a constant)

The gradient function of the cubic curve

We now return to the cubic function $y = x^3$.

We have these results:

x	gradient dy/dx
1	3
2	12
−3	27

We need to establish a relationship between x and dy/dx for the function $y = x^3$, and we see from the table that it is not a straight line. A moment's observation shows that dy/dx is divisible by 3 in each case, and we can rewrite the table as:

x	gradient dy/dx
1	3×1
2	3×4
−3	3×9

and we see that $dy/dx = 3x^2$. This is our second important result:

The gradient function of the cubic curve is a quadratic function.

Tutorial

Progress question (4)

This question summarises the results of this chapter.

Using the results from Practice Questions (4.2), complete this table:

y	dy/dx
x^3	
ax^2	
bx	
c	

Practical assignment (4)

Roll a tennis ball, or similar type of ball, along a stretch of floor. Choose a carpeted rather than a smooth floor. Observe how the speed of the ball changes during its short journey, and sketch a graph of its movement, using the x-axis for time and the y-axis for the **velocity** (not the distance) of the ball. Roll the ball again, varying the starting velocity by rolling it harder or more gently, and sketch a graph for each journey. Compare and interpret the gradients of your curves.

Seminar discussion

A curve represents a changing quantity, but not necessarily movement. Suggest quantities, other than time, distance and speed, to which the techniques of calculus might be applied.

> **Study tip**
> Look at the table in Progress Question (4) above, then rewrite it without reference to the book. Check that you know these results well before moving on.

How to differentiate algebraic functions

One-minute summary

In this chapter we extend the range of functions we can differentiate from the scant few in Chapter 4, to a surprisingly large set of algebraic functions. Before working through this chapter, check that you are confident with:
- multiplying out brackets
- simplifying easy algebraic expressions
- understanding negative and fractional powers

Revise these topics now if you are in any doubt. If you don't have a suitable book to hand, you could work through sections 4 and 5 of Chapter 14 – "Practising Algebra" and "Understanding Powers and Surds", or you could use these websites:
www.bbc.co.uk/schools/gcsebitesize/maths *or*
www.easymaths.com/problems_main.htm

Deductions from previous results

In Chapter 4 we obtained these gradient functions:

y	dy/dx
x^3	$3x^2$
ax^2	$2ax$
bx	b
c	0

Fig 5.1

Our next step is to find a general result which will:

- encompass these separate gradient functions and
- enable us to find the gradient function for any curve of the form $y = ax^3 + bx^2 + cx + d$.

Then we will extend this to include curves whose functions contain terms with higher powers of x.

Closer inspection of the table in Fig 5.1 shows the following relationships in each case:

(i) The power of x in dy/dx is 1 less than the power of x in the equation of the curve.

(ii) The multiplier in dy/dx is equal to the power of x in the equation of the curve. (Remember that $x^0 = 1$).

So, we can take these two steps to get from y to dy/dx:

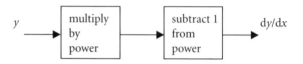

For the function $y = x^3$, the steps are:

$$y = x^3 \longrightarrow 3x^3 \longrightarrow dy/dx = 3x^2$$

Practice questions (5.1)
Apply the above process to these functions:
(a) $y = x^2$ (b) $y = 3x$
(c) $y = ax$ (a is constant) (d) $y = 5$
(e) $y = k$ (k is a constant) (f) $y = 4x^2$
(g) $y = 2x^3$ (h) $y = 0$
(i) $y = x^4$
Check your answers before moving on.
Now we can extend the range of functions which we can differentiate.

The general rule for $y = x^n$

The relationship we have discovered between y and dy/dx applies also to higher powers of x. The general rule can be written:

If $y = x^n$ then $dy/dx = nx^{n-1}$

This rule is of fundamental importance. You will need to use it again when you move on to more advanced differentiation, so **learn it now.**

The addition rule

It is a fortunate fact (which we are not going to prove here) that a function consisting of a sum of terms can be differentiated by:

- applying the process to each term separately, and
- adding the results.

We hinted at this in Practical Assignment (2) (see page 10)

Example 5.1

$$y = 2x^5 - x^4 + 4x^2 - 6$$
$$dy/dx = 10x^4 - 4x^3 + 8x \text{ (the last term is 0)}$$

Algebraic manipulation

There are algebraic functions which are not expressed as a series of terms, but which can be manipulated into this form using the rules of algebra, and can then be differentiated.

Example 5.2

$$y = (x + 5)$$

We can multiply out the bracket, and the function becomes:

$$y = x^2 + 10x + 25$$

Then we can differentiate:

$$dy/dx = 2x + 10$$

Example 5.3

$$y = \frac{x^2 - 4x}{2x}$$

We can divide each term of the numerator by $2x$:

$$y = \tfrac{1}{2}x - 2$$

Now we can differentiate:

$$dy/dx = \tfrac{1}{2}$$

Negative and fractional powers

Let's return to the function $y = 1/x$.

This function can be written $y = x^{-1}$. Applying the process as before gives us $dy/dx = -x^{-2}$ which we can write as $dy/dx = -1/x^2$ if we prefer. We can apply the differentiation process to any negative power.

If the power is a fraction, e.g. $y = x^{\frac{1}{2}}$, we can apply the same process, and here we get $dy/dx = \tfrac{1}{2}x^{-\frac{1}{2}}$.

Practice questions (5.2)

Differentiate these functions:

(a) $y = (x - 4)^2$ (b) $y = (2x + 1)^2$ (c) $y = (x + 2)^3$

(d) $y = \dfrac{2x^2 - 5x}{x}$ (e) $y = 1/x^2$ (f) $y = 2x^{\frac{1}{2}}$

(g) $y = x^{\frac{1}{4}}$ (h) $y = x^{-\frac{1}{2}}$ (i) $y = \dfrac{x^2 + 2x + 3}{x}$

(j) $y = \dfrac{x^2 + 2x}{2x}$ (k) $y = x^{-4}$ (l) $y = x^{\frac{3}{4}}$

(m) $y = (x^{\frac{1}{2}} - 1)^2$ (n) $y = \dfrac{x - x^{\frac{1}{2}}}{x}$ (o) $y = \sqrt{x}$

Tutorial

Progress questions (5)

1. (a) Find the co-ordinates of the point on the curve $y = 3 - x^2$ where the gradient $= -1$.

   Hints: Find dy/dx. Then form an equation by putting $dy/dx = -1$. Solve the equation for x. Put this value of x into the equation of the curve to find the y co-ordinate.

   (b) Find the co-ordinates of the point on the curve $y = 2x^2 + x + 1$ where the gradient $= 5$.

   (c) Find the co-ordinates of the 2 points on the curve $y = x^3$ where the gradient $= 1.5$

2. (a) Differentiate the function $y = (x + 2)^3$, as follows:
   First expand the bracket, and simplify to four terms.
   Then find dy/dx and factorise the result.
   Could you have taken a short cut?

   (b) Try the same process with the function $y = (3x - 1)^3$.
   Does the same short cut work this time?

3. Look back at Practice Questions (5.2), (a), (b), (c). Can you take a similar short cut in these examples?

Practical assignment (5)

(a) Sketch the curve $y = 1/x$. You should be able to do this without referring back.

(b) Describe how the gradient changes:
   (i) as x increases in the 1st quadrant ($x > 0$) and
   (ii) as x decreases in the 3rd quadrant ($x < 0$)

(c) Using your observations from (b), sketch the curve of the gradient function.

(d) Sketch the curve $y = -1/x^2$

(e) Comment on your results.

Seminar discussion

Before the development of calculus, the science of physics was limited to working with static (unchanging) quantities. Mathematical treatment of motion or other forms of change was not possible. If we were to remove from our lives all the inventions which have been made possible by knowledge of the mathematics of motion and other changing quantities, what sort of life style would we have?

> **Study tip**
> Make absolutely sure that you know the rule:

$$\text{If } y = x^n \text{ then } dy/dx = nx^{n-1}$$

6 How second order differentiation works

One-minute summary

"Second order" just means "do it again"! Differentiating a function twice enables us to find out more about the function. In this chapter we establish:

- the notation for second order differentiation
- the relationships between the gradient function and the second derivative

This leads to some very important techniques in Chapter 7.

Repeated differentiation

We have seen that when we differentiate a function, we get a new function, which represents the gradient of the original function. It is therefore possible to repeat this process, i.e. we can obtain the gradient function of the gradient function!

Example 6.1

$$y = x^3 \quad dy/dx = 3x^2$$

The function $dy/dx = 3x^2$ can be differentiated. But first we need a "label" for the new function. The notation for this is:

$$d^2y/dx^2$$

(This is pronounced as "d 2 y by d x squared" or "d squared y by d x squared")

In this case, $d^2y/dx^2 = 6x$.

Terminology

A function which is obtained from another function by differentiation is called a **derived function** or a **derivative**. These terms are more general than **gradient function**, and are used because in many applications of calculus, a function is differentiated for reasons other than finding a gradient.

These terms lead to the use of:

"first derivative" (i.e. dy/dx) and

"second derivative" (i.e. d^2y/dx^2)

Practice questions (6.1)

For the following functions, write down the functions dy/dx and d^2y / dx^2:

(a) $y = 3x^2 + 5x$ (b) $y = \frac{1}{2} x^3$ (c) $y = 2x^4 - x^3$

(d) $y = 2/x$ (e) $y = 1/x^2$ (f) $y = x$

What are maximum and minimum values?

In many applications of calculus, we are interested in the maximum (highest) or minimum (lowest) value of a function. These terms apply to the places where a curve reaches a peak (high point) or a trough (low point), rather than to the highest or lowest value of a function across the full range of values of x.

The function $y = x^2$ has its minimum value when $x = 0$, but the function $y = x^3$ has no minimum or maximum value in the sense we mean here, because the curve does not reach a high point and turn down again, or reach a low point and turn up again. However, if you consider just the range of values of x from, say -2 to $+2$, then the function has a least value of -8 and a greatest value of $+8$ in this range. These are not minimum or maximum values in the sense we mean in differentiation.

Tutorial

Progress questions (6)

1. Sketch the curve $y = x^2$ and answer these questions:
   (a) What are the co-ordinates of the point where the function has its minimum value?
   (b) What is the minimum value of this function?
   (c) Does this function have a maximum value?
   (d) What is the gradient of the function at the point where it reaches its minimum?
2. Sketch the curve $y = 2 - x^2$ and answer these questions:
   (a) What are the co-ordinates of the point where the function has its maximum value?
   (b) What is the maximum value of this function?
   (c) Does this function have a minimum value?
   (d) What is the gradient of the function at the point where it reaches its maximum?

Practical assignment (6)

Divide a sheet of graph paper horizontally into three sections. Draw x and y axes in each section, and place the y-axis in the same central position through all three sections (so that your graphs will line up one above the other).

In the top section, sketch the curve $y = x^2$.
In the middle section, sketch its gradient function, dy/dx.
In the lower section, sketch the gradient of the gradient, d^2y/dx^2

From your sketches, answer these questions:

1. (a) Describe the gradient of the curve near the origin:
       (i) where $x < 0$
       (ii) where $x > 0$
   (b) What is the gradient of the gradient at the origin?
2. Repeat this exercise for the curve $y = -x^2$.
3. Repeat this exercise for the curve $y = x^3$.

Seminar discussion

If the variable t represents time, and the variable s represents distance, what does the function ds/dt represent? What does the function d^2s/dt^2 represent? Would the function d^3s/dt^3 have any meaning, and if so, what?

> ### Study tip
> Further practice in looking at gradients, and gradients of gradients, would be very helpful at this stage. Work through Practical Assignment (6) again. For further practice, investigate the curve $y = -x^3$ in the same way.

7 Stationary points explained

One-minute summary

"Stationary" means "standing still". "Stationary points" are points on a curve where the function is neither increasing nor decreasing, so the gradient is zero at these points. This has great significance in many applications of calculus, so it is very important for mathematicians to be able to find the stationary points on a curve. In this chapter we will:

- see what stationary points look like
- use differentiation and second order differentiation to find and analyse stationary points.

What do stationary points look like?

A point on a curve where the gradient is zero is called a **stationary point**. There are 3 types of stationary point which we need to explore:

(i) A maximum point is where the gradient changes from positive, to zero at the maximum, and then becomes negative.

(ii) A minimum point is where the gradient changes from negative, to zero at the minimum, and then becomes positive.

(iii) A point of inflexion is **either** where the gradient changes from positive, through zero at the point of inflexion, and then becomes positive again, **or** where the gradient changes from negative, to zero at the point of inflexion, and then becomes negative again.

These are illustrated below.

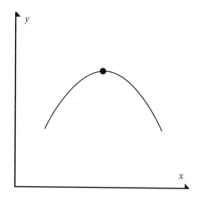

Fig 7.1   A maximum point

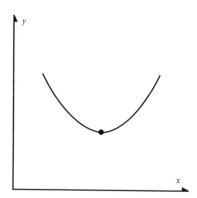

Fig 7.2   A minimum point

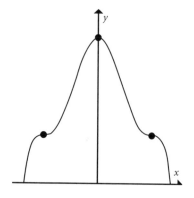

Fig 7.3   Two points of inflexion and a maximum point

[Note: There is another type of point of inflexion which does not have zero gradient. We do not need to include this here.]

Practice questions (7.1)

1. Name a function which has a point of inflexion. Is the gradient positive or negative on each side of its point of inflexion?
2. Use your answer to question 1 to find a function which has the other type of point of inflexion.
3. Name a function which has none of the stationary points shown in Figs 7.1 to 7.3.
4. Sketch the curve $y = x^3 - 3x$ and describe its stationary points.

Using differentiation to find stationary points

We can use differentiation to find the maximum or minimum value of a function.

We will use the curve $y = x^3 - 3x$ to verify the results from question 4 above.

First, we find the gradient function:

$$dy/dx = 3x^2 - 3$$

Since we are looking for a point where the gradient is zero, we need to equate dy/dx to zero:

$$3x^2 - 3 = 0$$

There are two solutions to this equation, $x = 1$ and $x = -1$

Putting these values back into $y = x^3 - 3x$ gives the two points $(1, -2)$ and $(-1, 2)$.

But how do we know which of these points is a maximum and which is a minimum? Sketching the curve is not a mathematically exact way of answering this question.

Distinguishing between maximum and minimum points

We know that at a **maximum** point, the gradient changes from positive to zero to negative. This means that the gradient function is **decreasing** at a maximum point. If the gradient function is decreasing, then **its** gradient function, i.e. d^2y/dx^2, will be negative.

We know that at a **minimum** point, the gradient changes from negative to zero to positive. This means that at the gradient function is **increasing** at a minimum point. If the gradient function is increasing, then **its** gradient function, i.e. d^2y/dx^2, will be positive.

This gives us a mathematical method for distinguishing between maximum and minimum points.

To distinguish between maximum and minimum points:

(i)   Find d^2y/dx^2
(ii) Evaluate d^2y/dx^2 at the stationary point
(iii) If $d^2y/dx^2 > 0$, the point is a minimum.
      If $d^2y/dx^2 < 0$, the point is a maximum.

Tutorial

Progress questions (7)

1. Examine the point of inflexion of the curves

    (i) $y = x^3$
    (ii) $y = -x^3$

    In each case, what can you say about the gradient function? What can you deduce about d^2y/dx^2 at the point of inflexion?

2. Use differentiation to find the stationary points of the function $y = x^3 + 3x^2 - 9x - 1$.

    State the co-ordinates of each stationary point and use the method shown in this chapter to establish which is a maximum and which is a minimum. Use your results to sketch the curve.

Practical assignment (7)

Cut out a rectangular piece of card measuring 20 cm by 12 cm.

Cut out a square from each corner measuring 1 cm by 1 cm.

Fold the edges of the card up to make a box without a lid.

Calculate the volume of the box in cm^3.

Then calculate what the volume of the box would be if you had cut off corners measuring 2 cm by 2 cm.

Try other sizes for the cut-off corners.

What is the biggest volume you can get for the box?

Method: Let the size of the cut-off corners be x cm by x cm. Write down an expression for the volume of the box. When you have expanded the brackets you will have a term in x^3, a term in x^2 and a term in x. Differentiate this expression, equate to zero, and solve the resulting quadratic equation. Check that one of the solutions is near to what you are expecting, and discard the other value. The correct value of x gives the maximum value for the box. Confirm that it is a maximum by obtaining the second derivative and substituting x (you should get a negative result). Lastly, find the volume of the box for this value of x.

Seminar discussion

Suggest other problems, similar to this practical assignment, which you could solve using differentiation. Why is calculus so much better than trial and improvement?

Study tip

Write out the rule for using $\mathrm{d}^2y/\mathrm{d}x^2$ to determine the nature of stationary points. Extend it to include points of inflexion (see the answers to Progress Questions (7)). Learn this rule.

8 How to apply the theory

One-minute summary

The Practical Assignment in Chapter 6 is an illustration of how differentiation can be used to solve problems for which we would otherwise have to rely on the laborious process of trial and improvement. Using calculus is much quicker, but the really important point is that we have the satisfaction of knowing for certain that our solution is the maximum or minimum value which we require (assuming that we haven't made a mistake in the mathematics!) In this Chapter we see how to apply differentiation to problems such as:

- minimising the length of fencing needed for an enclosure
- minimising the amount of material needed to enclose a fixed volume

Areas, perimeters and volumes

Before continuing with this Chapter, read again through the solution to Practical Assignment (7).

This cardboard box exercise gives an example of the use of differentiation to find the maximum volume of a container, and is a simple illustration of how calculus can be used very effectively in solving simple design problems.

This chapter provides further examples of applying the technique of differentiation to practical geometry.

Example 8.1
Find the lengths of the sides of a rectangular enclosure, given that 100 metres of fencing is available, and the area of the enclosure is to be maximised.

If we could not apply calculus to this problem, we would, as before, write down a table of possible dimensions for the enclosure and work out the area in each case:

Length	Width	Area
40	10	400
35	15	525
30	20	600
25	25	625

This suggests that a square is the "best" rectangle from the point of view of enclosing the maximum possible space.

Now let's confirm this by differentiation:

Let x be the length of the rectangle, then the width $= 50 - x$.

Then the area $A = x(50 - x)$ and this is the function we want to maximise.

$$A = 50x - x^2$$
$$dA/dx = 50 - 2x$$

For maximum area, $dA/dx = 0$,

i.e. $50 - 2x = 0$

So $x = 25$ and $A = 25 \times 25 = 625$.

Example 8.2

Following on from Example 8.1, suppose a wall is available to provide one side for the enclosure. What is the maximum area we can now enclose with 100 metres of fencing?

The fencing will provide three sides of the enclosure. (We are assuming that the wall is longer than the enclosure's longer side).

Let $x =$ length (parallel to the wall).

Then width $= (100 - x)/2 = 50 - \frac{1}{2}x$

Area, $A = x(50 - \frac{1}{2}x) = 50x - \frac{1}{2}x^2$

To maximise the area, find dA/dx and equate it to zero:

$$dA/dx = 50 - x = 0$$
$$x = 50$$

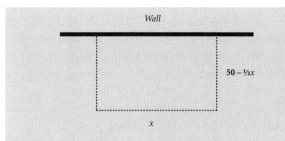

Fig 8.1

So the length of the enclosure will be 50, the width is 25, and the area is $50 \times 25 = 1250$.

This time a square enclosure is not the best option. Why is this?

If we made a square enclosure, each side would be of length $100/3 = 33.33$ (to 2 d.p.), and the area would be $33.33 \times 33.33 = 1110.89$ (to 2 d.p.). The reason this provides a smaller area is because we would not be making best use of the "free" wall – we would be using 33.33 metres of wall, but the best solution uses 50 metres of wall.

Practice questions (8.1)

1. Two walls at right angles can be used to form part of a rectangular enclosure. One of the walls is 30 metres long, and the other is 100 metres long. How should the 100 metres of fencing be arranged to maximise the enclosed area? What will the maximum area be?

2. A box without a lid is to be made from a square piece of cardboard measuring 10 cm by 10 cm, by cutting squares from each corner and folding the sides up. What size should the cut-off squares be for the box to contain the maximum volume? What will the maximum volume be?

Problems involving more than one variable

Our ability to solve the problems we have looked at so far, has depended on the fact that there has been only one variable to consider – the length of the side of a rectangle, the size of the cut-off corners of the box, etc.

The technique of differentiation which we are using is simple differentiation where y is a function of **one** variable x.

But we can tackle problems with two variables provided that the two variables are related mathematically, and that we can express one in terms of the other. If we can do this, then the problem reduces to a function of one variable and we can proceed as before.

We have in fact already seen one such example. In Practice Questions (8.1) 1, the extension to the 30 metre wall, e, was expressed in terms of x by using the fixed total length of the fencing. The area was then written as a function of x only and we were able to differentiate it.

Example 8.3

A manufacturer of canned drinks is finding that costs are increasing because of the rising prices of aluminium. They need to reduce the amount of aluminium used without reducing the quantity of drink in the cans.

They are currently using cylindrical cans with a diameter of 6 cm. The capacity of the cans is 250 ml. Is it possible to reduce the amount of aluminium used?

To solve this problem, we need to find an expression for the amount of aluminium used – this is directly related to the surface area of the can. But there are two variables – the height and the diameter. How are these two variables related?

The key is the volume of the can, which is directly related to the capacity. One millilitre (ml) occupies 1 cm³, so the volume of the can is 250 cm³.

We know that the volume of a cylinder is given by the formula:

$$V = \pi r^2 h$$

where r is the radius and h is the height.

But V is a constant; $V = 250$ cm³

So now we can write $\pi r^2 h = 250$

and express h in terms of r:

$$h = 250/\pi r^2$$

Now we can look at the area of aluminium used to make the can. This is given by the formula:

$$A = 2\pi r^2 + 2\pi rh$$

(where the first term is the top and base of the can and the second term is the curved surface).

This is the function we want to minimise. In this form, we cannot differentiate it because h and r are both variables – if one changes, so does the other – because the volume remains constant.

Our next step is to write A as a function of r only, by substituting $h = 250/\pi r^2$ and simplifying:

$$A = 2\pi r^2 + 500/r$$

Now we can differentiate:

$$dA/dr = 4\pi r - 500/r^2$$

Putting $dA/dr = 0$ gives $r^3 = 500/4\pi$ i.e.

$$r = 3.414 \quad \text{(to 3 d.p.)}$$

The drinks company have been using cans with diameter = 6 cm, i.e. $r = 3$. So they can save on aluminium by using cans which are wider (and hence shorter). See Progress Questions (8) for the conclusion of this problem.

Note that although none of the steps in this working is difficult, it is very important to spot the fact that there are two variables.

If we had made what is a very common error, by differentiating A with respect to r before substituting for h, we would have been treating h as a constant – and it would have been impossible to arrive at a solution.

(In higher mathematics, there are methods of differentiation for functions of more than one variable. These methods are needed when the variables are independent. Such mathematics is well outside the scope of this book.)

Tutorial

Progress questions (8)

Complete the drinks can problem as follows:

a) Find the height of the can and the area of the aluminium when the radius was 3 cm.
b) Find the height of the can and the area of the aluminium if the radius is changed to 3.4 cm.
c) Assuming that the aluminium used is 0.5 mm thick, find the percentage saving in the volume of aluminium used if the radius is changed to 3.4 cm.

Practical assignment (8)

An open-topped water tank with a square base has a capacity of 4000 litres. Find the height and the width of the tank. Then find the area of the metal, i.e. the surface area of the tank.

Seminar discussion

Why might it not always be possible to put into practice a mathematical solution to a problem, such as changing the shape of the drinks cans to reduce the production costs?

Study tip

Differentiation is just one of several steps which are required in finding the solutions to the types of problems we have seen in this chapter. Common errors in tackling these problems are errors of omission – forgetting that the solution is not complete until all the dimensions and the maximum or minimum value has been found. Read through the drinks can problem again and notice how many calculations need to be done to complete the whole problem – including the parts in Progress Questions (8).

9 Understanding tangents and normals

One-minute summary

We return to pure mathematics for an extension of our previous work on gradients. In this chapter we use two things which you have probably learnt from co-ordinate geometry:

- the formula for the equation of a straight line
- the relationship between the gradients of perpendicular lines

If you need to revise these, work through section 6 "Basic Co-ordinate Geometry" in Chapter 14.

You could also look at:

http://www.mathsnet.net/asa2/modules/p1.html#4

Tangents and normals are important in particular areas of Co-ordinate Geometry such as conic sections.

Equation of a tangent

We mentioned tangents in Chapter 3 when we started to look at the gradient of a curve. A tangent is a straight line which touches a curve, and has the same gradient as the curve at the point of contact.

Now that we are able to find the gradient of a curve at a given point, we can use our knowledge of co-ordinate geometry to find the equation of the tangent to the curve at that point.

Example 9.1

Find the equation of the tangent which touches the curve $y = x^2$ at the point A $(1,1)$

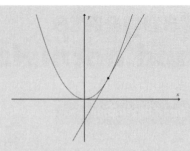

Fig 9.1

The gradient function is $dy/dx = 2x$. At A, $dy/dx = 2$.

The equation of a straight line, $y = mx + c$, can be found from its gradient and one point on the line.

We have $m = 2$ and the point $(1,1)$.

The formula we use (from Co-ordinate Geometry) is:

$$y - y_1 = m(x - x_1)$$

This gives

$$y - 1 = 2(x - 1)$$

simplifying to

$$y = 2x - 1$$

This is the equation of the tangent to the curve $y = x^2$ at the point A $(1,1)$

[Note: If you are not familiar with this method of finding the equation of a straight line, you could work through the section "Basic Co-ordinate Geometry" in the Appendix on pages 103–106.]

Practice questions (9.1)

1. Find the equation of the tangent to the curve $y = x^2$ at the point P $(-2,4)$

2. Find the equation of the tangent to the curve $y = x^3$ at the point Q $(0.5, 0.125)$

3. If the line $y = mx + c$ is a tangent to the curve $y = x^2 + c$, what is the value of m?

You do not need to use differentiation to answer this question – you can deduce the answer from a sketch.

Equation of a normal

A normal is a straight line which intersects the tangent at right angles, at the point of contact of the tangent and the curve.

Example 9.2

Find the equation of the normal to the curve $y = -x^2 + 2x + 3$ at the point (2,3).

First, we find the gradient of the curve at the point (2,3):

$$\mathrm{d}y/\mathrm{d}x = 2x + 2$$

When $x = 2$, $\mathrm{d}y/\mathrm{d}x = -2$.

The normal is perpendicular to the tangent, and we know from Co-ordinate Geometry that if gradients m_1 and m_2 are perpendicular, then

$$m_1 \times m_2 = -1$$

Using m_1 for the tangent and m_2 for the normal, we have $m_1 = -2$ and $m_2 = \frac{1}{2}$

We have $x_1 = 2$ and $y_1 = 3$, so using the straight line formula:

$$y - y_1 = m(x - x_1)$$

we get

$$y - 3 = \frac{1}{2}(x - 2)$$

which simplifies to

$$y = \frac{1}{2}x + 2$$

This is the equation of the normal to the curve $y = -x^2 + 2x + 3$ at the point (2,3).

The equation of the tangent is left for you to do (in Practice Questions (9.2)).

Fig 9.2 shows the curve $y = -x^2 + 2x + 3$ and its tangent and normal at the point (2,3).

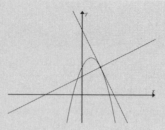

Fig 9.2

Practice questions (9.2)

1. Find the equation of the tangent in Example 9.2.
2. Find the equations of the normals to the curve $y = x^2$:
   (i) at the point A $(1,1)$.
   (ii) at the point B $(-2,4)$.
3. Find the equation of the normal to the curve $y = x^3$ at the point Q $(0.5, 0.125)$.
4. (i) At which two points on the curve $y = 1/x$ is the line $y = x$ a normal to the curve?
   (ii) What are the equations of the two corresponding tangents?
5. (i) Write down the co-ordinates of the points where the curve $y = x(x - 1)(x + 1)$ cuts the x-axis.
   (ii) Find the gradient function of this curve, and the value of the gradient at the origin.
   (iii) Use the results of (i) and (ii) to sketch the curve.
   (iv) Deduce the equation of the normal to the curve at the origin.
   (v) Find the co-ordinates of the points P and R where the normal meets the curve again.
   (vi) Find the gradient of the curve at P, and deduce the gradient of the curve at R.

Tutorial

Progress questions (9)

1. The tangent to the curve $y = x^2$ at the point A $(1,1)$ meets the y-axis at the point M.
   The normal to the curve at A meets the y-axis at the point N.
   (i)   Find the co-ordinates of the points M and N.
   (ii) Find the distance MN.
2. The line $y = x + 3$ is a tangent to the curve $y = x^2 + a$ at the point P.
   (i)   Find the co-ordinates of P.
   (ii) Find the value of a.
   (iii) Find the equation of the normal to the curve at P.
3. The line $y = -3x$ is a normal to the curve $y = k - x^2$ at the point N.
   (i)   Find the co-ordinates of N.
   (ii) Find the value of k.
   (iii) Find the equation of the tangent to the curve at N.

Practical assignment (9)

Use your graphics calculator to find the co-ordinates of the point where the normal to the curve $y = x^2$ at the point $(1,1)$ intersects with the curve in the 2nd quadrant. Alternatively, find the equation of the normal, form a quadratic equation, solve it for x and lastly find the y co-ordinate.

Seminar discussion

Can you find an example of a tangent intersecting the curve at the point of contact?

Study tip
Revise the rules of Co-ordinate Geometry which have been used in this chapter:

(i) the formula for the equation of a straight line
(ii) the rule for perpendicular gradients

The chain rule explained

In this chapter we extend the range of functions we can differentiate, to include functions which are differentiated in two steps rather than one.

The reason for the use of the term "chain rule" will become apparent when you see how we differentiate these functions.

Two-step differentiation

Look again at Practice Questions (5.2). In Question (a) we differentiated the function $y = (x - 4)^2$ as follows:

Expand the bracket:

$$y = x^2 - 8x + 16$$

Differentiate:

$$dy/dx = 2x - 8$$

Factorise:

$$dy/dx = 2(x - 4)$$

We see from this result that we could have applied the basic rule $dy/dx = nx^{n-1}$ with n = 2, and $(x - 4)$ in place of x. This cuts out the need to expand the bracket.

In Question (b), applying the full method, we have:

$$y = (2x + 1)^2$$

Expand the bracket:

$$y = 4x^2 + 4x + 1$$

Differentiate:

$$dy/dx = 8x + 4$$

Factorise:

$$dy/dx = 4(2x + 1)$$

If we had used the short method in this case, with $(2x + 1)$ in place of x, we would have arrived at $dy/dx = 2(2x + 1)$ which is wrong – a factor of 2 is missing. This missing factor is the derivative of the function $(2x + 1)$. In question (a) the function is $(x - 4)$, and the derivative of this function is 1, so nothing is lost when we use the short method.

But now that we know that we need to include this factor in the differentiation, we can build it into the new method.

This method of differentiation is known as the "chain rule" or the "function of a function" rule. Both of these names indicate that the method involve more than one step in the differentiation. Not only does this method remove the need to expand the bracket in the examples we have seen so far, but it also brings more functions into the range of those which we can differentiate.

Example 10.1

$$y = (x^2 - 1)$$

It would be extremely laborious to differentiate this function by expanding the bracket. But with the chain rule we work as follows:

$$dy/dx = 9(x^2 - 1)^8 \times (2x)$$

where the factor $2x$ is the derivative of the function $(x^2 - 1)$

We then write this a bit more tidily as

$$dy/dx = 18x(x^2 - 1)^8$$

Example 10.2

$$y = (2x + 3)^{1/2}$$

In this case, we can only differentiate this function by using the chain rule, because expanding the bracket would give an infinite series.

$$dy/dx = \tfrac{1}{2}(2x + 3)^{-1/2} \times 2$$

where the factor 2 is the derivative of the function $(2x + 3)$
Then we tidy up:

$$dy/dx = (2x + 3)^{-1/2}$$

Practice questions (10.1)

Differentiate these functions using the chain rule:

(a) $y = (x - 2)^6$ (b) $y = (x^2 + 1)^2$

(c) $y = (x^2 + 2x)^{-1}$ (d) $y = (x^2 - 3x)^4$

(e) $y = (1 - x^{1/2})^3$ (f) $y = (x + 1/x)^2$

(g) $y = (1 + 2x)^{1/2}$ (h) $y = (x^3 - x)^{-3}$

(i) $y = (\tfrac{1}{2}x^2 + 1)^3$ (j) $y = (2x - 3)^4$

Check your answers before you move on.

We can write the chain rule like this:

If y is a function of x, then the derivative of y^n is
$$ny^{n-1} \times dy/dx$$

We can also use the chain rule in this form, if a function is written in the form $y^n = f(x)$.

Example 10.3

$$y^2 = 3x^2 + 1$$

Applying the chain rule to the left hand side gives:

$$2y\frac{dy}{dx} = 6x$$

which can be rearranged:

$$\frac{dy}{dx} = \frac{6x}{2y} = \frac{3x}{y}$$

So using the chain rule in this form gives dy/dx in terms of x and y.

If we want dy/dx in terms of x only, then we can use the original equation to write y in terms of x:

$$y = (3x^2 + 1)^{\frac{1}{2}}$$

and then substitute for y in the expression for dy/dx:

$$\frac{dy}{dx} = \frac{3x}{(3x^2 + 1)^{\frac{1}{2}}}$$

We could have got this same result by starting with $y = (3x^2 + 1)^{\frac{1}{2}}$. Either approach is equally good.

But sometimes it is easier to use the chain rule this way. And later on you will meet functions which you cannot write in the form $y = f(x)$, so this method will be important then.

Practice questions (10.2)

Use the method shown in Example 10.3 to differentiate these functions.

Leave your answers in terms of x and y. In (d) and (e) don't re-arrange the equation, just differentiate each term.

(a) $y^2 = 2x^2 - x + 1$ (b) $y^2 = 4x - 1$

(c) $y^3 = 3x + 2$ (d) $x^2 + y^2 = 1$

(e) $y^3 + 1 = x$ (f) $y^2 = x^{1/2}$

(g) $y^4 = x$ (h) $y^{-1} = x^3$

Tutorial

Progress questions (10)

1. (a) Use the chain rule to differentiate the function
   $y = (x^2 + 2x + 1)^2$
   (b) Factorise fully your answer to (a)
   (c) Factorise the function in (a) and use the chain rule to differentiate it in its factorised form.
   (d) Check that your answers to (b) and (d) are the same.

2. (a) Use the chain rule (twice) to differentiate the function $y = ((x + 1)2 - 1)2$. Simplify and factorise your answer.
   (b) Simplify the function in (a) by expanding the brackets and reducing to an expression containing three terms. Differentiate this function and check that your answers to (a) and (b) agree.

3. (a) In Practice Questions (10.2), take each of the functions in (g) and (h). Express the function in terms of y and differentiate. Check that your answers are the same as before. Your previous answers will be in terms of y and x, so you will have to express them in terms of x only before comparing.

Practical assignment

Write a question (and its solution!) using the Chain Rule which would be suitable to add to Progress Questions (10). Give the question to a fellow student and check their solution against your own. Alternatively, use your question for your own revision later.

Seminar discussion

What assumptions are we making when we apply the chain rule?

> **Study tip**
> Look back through the book and write down all the rules of differentiation which you have learned so far. Then write them out again from memory.

Understanding integration

One-minute summary

The importance of integration is in its applications to mathematical problems. But first we need to learn how to integrate. In this chapter we will:
- revise basic differentiation
- see how to do the process in reverse
- introduce the notation for integration

Then we give you plenty of practice in basic integration so that you will be really familiar with the process when we apply it to problems in Chapter 12.

Reversing the differentiation process

The second main tool in calculus is the process of Integration. This is the opposite of differentiation. Given a gradient function, can we find the function from which it was derived?

The general answer to this question is "in some cases". However, in this book we are concerned only with simple integration where it **will** be possible to reverse the differentiation process.

First, we'll do a bit of revision of gradient functions.

Practice questions (11.1)
1. Find the gradient functions for these curves:
   (a) $y = 3x^2$ (b) $y = 3x^2 + 1$ (c) $y = 3x^2 - 4$
2. Write down another function which shares the same gradient function as the examples given in question 1.

3. Find the gradient functions for these curves:
   (a) $y = 2x^2 - 1$ (b) $y = 2x^2 + 1$ (c) $y = 2x^2 - 14$
4. Write down another function which shares the same gradient function as the examples given in question 3.

Differentiation – a many-one mapping explained

We see from the answers to these questions that the process of differentiation is "many-one". There are **many** (in fact an infinite number of) functions which all have the **one** gradient function $dy/dx = 6x$.

How can we reverse the process of differentiation if there are infinitely many answers?

Finding the original functions from the gradient function

We deal with this by introducing the idea of the "arbitrary constant". "Arbitrary" means "able to take any value". We generally use the letter c for an arbitrary constant, though this is only a convention – you can use any letter you like.

So we say that if $dy/dx = 6x$, then $y = 3x^2 + c$. c can be negative or positive or zero, so this answer covers all possibilities.

Now all we need is the notation for integration, and you can do your first integrals.

There is a special sign in mathematics to indicate integration. It looks like this: $\int$

We write the integral sign, followed by the function we want to integrate, followed by "dx" to indicate that x is the variable.

The example above would be notated as follows:

$$\int 6x \, dx = 3x^2 + c$$

If the function to be integrated has more than one term, it should be enclosed in brackets:

$$\int (2x + 1)\, dx = x^2 + x + c$$

Check this example is correct by differentiating $y = x^2 + x + c$.

Practice questions (11.2)
Find the following integrals:

(a) $\int 12x\, dx$ (b) $\int (2x-3)\, dx$ (c) $\int (4x+2)\, dx$

Check your answers by differentiation – if you arrive back at the function to be integrated, then your answer is correct.

Checking back by differentiation should always be part of the integration process – it isn't just for beginners.

Practice questions (11.3)

(a) $\int 3x^2\, dx$ (b) $\int x^2\, dx$ (c) $\int 4x^3\, dx$

(d) $\int x^3\, dx$ (e) $\int -x^{-2}\, dx$ (f) $\int 2x^{-2}\, dx$

(g) $\int nx^{n-1}\, dx$ (h) $\int x^n\, dx$

Check your answers to these questions before moving on.

The general rule for $dy/dx = x^n$

In working through this last set of integrals, you will have noticed that you can apply the opposite process to the general rule for differentiation.

In Chapter 4, we found this two-step process for differentiation:

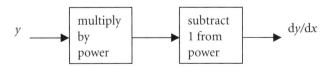

Now we can see that the reverse of this process enables us to integrate functions which consist of terms in powers of x:

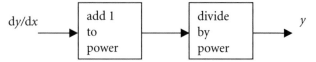

We can express this rule algebraically:

$$\text{If } dy/dx = x^n \text{ then } y = \frac{x^{n+1}}{n+1}$$

Tutorial

Progress questions (11)

Here is a mixed set of integrals. You may need to refer back to Chapter 5, and in some cases you will need to manipulate the given function to get it into a form which you can integrate.

(a) $\int 6x^8 \, dx$ (b) $\int (x^2 - 1) \, dx$ (c) $\int 4 \, dx$

(d) $\int (x^{-3} - 3) \, dx$ (e) $\int \dfrac{(1-x^2)}{x^2} \, dx$ (f) $\int (x+2)^2 \, dx$

(g) $\int x^{2n} \, dx$ (h) $\int x^{1/2} \, dx$

Practical assignment (11)

(a) Find the area of a rectangle bounded by the axes, the line $y = a$ (a is a constant) and the vertical line through a general point on the x-axis, $(x,0)$.

(b) Find the area of a triangle bounded by the x-axis, the line $y = x$, and the vertical line through a general point on the x-axis, $(x,0)$.

(c) Comment on your results.

Seminar discussion

Is there a value of n for which the rule for the integration of x^n is not valid?

> **Study tip**
> Learn the rule for the integration of x^n

12 Understanding the area under a curve

One-minute summary

Integration is a powerful tool which can be used to find areas and volumes of shapes whose boundaries are not straight lines. If the boundary is a curve which we can express as a function, then we can integrate the function and hence find the area between the curve and the x-axis (or y-axis).

Essentially, integration involves splitting an area into an infinitely large number of infinitely thin strips, finding the area of each strip and adding these areas together. In this chapter we see how to:

- find the area of a shape bounded by a curve and the x-axis over a given range of values of x
- find the area of a shape bounded by a curve and the y-axis over a given range of values of y

Indefinite integrals

In Chapter 11, we looked at the process of integration as the reverse of differentiation, and the answers to the integrals were functions of x. Each such function, e.g. $y = 3x + c$, represents an infinite set of curves (or straight lines) because the constant c can take any value. An integral whose answer is a **function** rather than a **numerical result** is called an indefinite integral.

Definite integrals

We now come to an important practical application of integration – finding the area under a curve. "Under a curve" in this context means the area between a curve and the x-axis, within a given range of values of x. So we will be looking at finite shapes, such as those in the Practical Assignment in Chapter 11, and the answers to our integrals will be numerical values. An integral whose answer is numerical is called a **definite integral**.

The approximate area under a curve

We can find the area under a curve by using approximate methods such as the Trapezium rule. These methods (which are not covered in this book) involve dividing the area into a number of simple shapes, and adding together the areas of these shapes.

Here is a diagram showing the area under a curve divided into rectangles of equal width.

We can add the areas of the rectangles together, and the resulting total area will be quite close to the actual area under the curve.

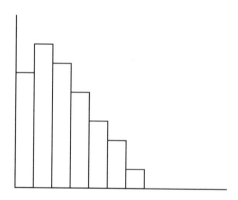

Fig 12.1

The height of each rectangle is given by y, i.e. the value of the function at the centre of the top of the rectangle. We can use dx for the width of each rectangle, as this represents the difference between the x – values at the bases of the rectangles. The area of each rectangle is then $(y \times dx)$ and the total area is:

$$\Sigma \, (y \times dx)$$

By making the rectangles narrower, we can obtain a result closer to the actual area under the curve:

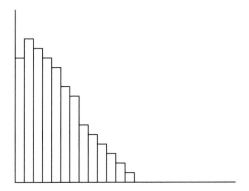

Fig 12.2

Now we apply a limiting process, similar to that used when we reduced the length of the chord AB in order to find the gradient of the curve at the point A – see Chapter 3 if you need to remind yourself about this. We choose a smaller value for dx. The smaller the value of dx, the greater the number of rectangles, the smaller their width, and the closer the total area is to the exact value we are seeking.

In the limit, we are adding an infinite number of rectangles of infinitesimally small width. Although we cannot do this arithmetically, the process of integration provides the method which gives an exact answer for the required area.

Notation: We replace the Σ (summation) sign by the $\int$ (integral) sign, we drop the multiplication sign, and we write:

$$\text{Area} = \int y \, dx$$

[Note: The real mathematics we are using here derives from a theorem called the Fundamental Theorem of Calculus. Such mathematics is well beyond the scope not only of this book but almost certainly of the course you are studying, but if you continue your study of this subject you will meet it in the future.]

Using integration to find areas

Example 12.1
Find the area between the x-axis, the lines $x = 0$ and $x = 4$, and the function $y = 3$.

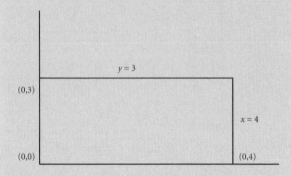

Fig 12.3

The obvious way to find this area is simple multiplication of the length and width of the rectangle:

$$4 \times 3 = 12$$

We want to see that we get the same result using integration.

The range of values of x over which we are integrating is $x = 0$ to $x = 4$. We call these values "limits" and write them at the bottom and the top of the integration sign. (The use of the word "limits" has nothing to do with "taking limits" but just means the ends of the range of values of x).

We write

$$\text{Area} = \int_0^4 y \, dx = \int_0^4 3 \, dx = \left[3x \right]_0^4$$

The function we have found by integration, $3x$, represents the area under the line $y = 3$, and to the left of the vertical line through the general point on the x-axis, $(x,0)$. The area of the rectangle is the area to the left of the vertical line through (4,0) **minus** the area to the left of the vertical line through (0,0). So we evaluate the function $3x$ at $x = 4$ (the upper limit) and at $x = 0$ (the lower limit) and take the difference:

$$\text{Area} = 3 \times 4 - 3 \times 0 = 12 \text{ sq units}$$

What happened to the constant, c?
Shouldn't we have written $\int 3 \, dx = 3x + c$?
Strictly speaking, yes. But we would then have written:

$$\text{Area} = (3 \times 4 + c) - (3 \times 0 + c) = 12 + c - c = 12 \text{ sq units}$$

A moment's thought will satisfy you that c will always disappear when we are integrating between limits. So we can allow ourselves the licence of omitting c when we are calculating definite integrals.

Example 12.2

Find the area between the curve $y = 3x^2$ and the x-axis, in the range $0 < x < 2$.
The area is given by $\int_{0}^{2} 3x^2 \, dx = \left[x^2 \right]_{0}^{2} = 8 - 0 = 8 \text{ sq units}$.

Practice questions (12)

1. Use integration to find the areas of:
   (i) the rectangle bounded by the lines $y = 3$, $x = 1$ and $x = 4$.
   (ii) the triangle bounded by the x-axis, the line $y = x$, and the line $x = 5$. Check your result using the formula for the area of a triangle.
   (iii) the trapezium bounded by the x-axis and the lines $y = 2x$, $x = 3$ and $x = 6$. Check your result using the formula for the area of a trapezium: ½ $(a + b) \times$ h
2. (i) Use the formula ½$(a + b) \times$ h to find the area of the trapezium bounded by the x-axis, the

line $x = 1$, the line $x = 2$, and the line joining the points (1,1) and (2,4).

(ii) Use integration to find the area of the shape bounded by the curve $y = x^2$, the x-axis, the line $x = 1$ and the line $x = 2$.
Give your answer correct to 2 d.p.

(iii) Which is the larger of these two areas and why?

Calculation of areas between a curve and the y-axis

Integration can also be used to find the area between a curve and the y-axis, for a range of values of y.

Example 12.3

Look again at Fig 12.3 (in Example 12.1).
We can find the area of the same rectangle, by integrating the function $x = 4$ between the limits $y = 0$ and $y = 3$. So we are looking at the area "under" (i.e. to the left of) the line $x = 4$, bounded by the y-axis.

We write
$$\text{Area} = \int_0^3 x \, dy = \int_0^3 4 \, dy = \left[4y \right]_0^3$$

Substituting the values 3 and 0 for y, and subtracting, gives:

$$\text{Area} = 4 \times 3 - 4 \times 0 = 12 \text{ sq units}$$

You will see that this process is exactly as before but with x and y reversed. For this trivially simple example, it makes no difference which way we do the integration. But there are sometimes cases where a solution can be found much more easily by integrating x as a function of y, and using the y-axis as a boundary. So it is important to know how to integrate between a curve and the y-axis.

Tutorial

Progress questions (12)

1. Find the area bounded by the curve $y = 2 - x^2$ and the x-axis. (In order to determine the limits of integration, you will first need to find the co-ordinates of the points where the curve cuts the x-axis). Give your answer correct to 2 d.p.

2. (a) Sketch the curve $y = x^{1/2}$ in the range $x = 0$ to $x = 4$. Take positive values of y only.
   Find the area between the curve, the x-axis, and the lines $x = 0$ and $x = 4$.
   Give your answer correct to 2 d.p.

   (b) In your sketch for 2(a), draw the line $y = 2$ to complete the rectangle. Find the area of the rectangle and subtract the area found in 2(a).

   (c) Confirm your answer to 2(b) by expressing x as a function of y in 2(a) and integrating this function with respect to y, using the y-axis as a boundary and appropriate limits for y.

3. Sketch the curve $y = 1/x^2$ in the range $0 < x < 2$. Find the area between the curve, the x-axis, and the lines $x = 1$ and $x = 2$.

Practical assignment (12)

Plot the curve $y = 2 - x^2$ from $x = 0$ to $x = 1.5$. Use graph paper, not squared paper.

Estimate the area between the curve and the x-axis by counting up the small squares.

How close is your answer to the exact answer given by integration in Progress questions (12) 1?

Seminar discussion

Is it possible to find the area between the curve $y = 1/x$ and the x- and y-axis?

Study tip

Chapter 13 contains practice questions which cover all the material in this book. Many of these Questions are more demanding than those you have met in earlier chapters. So this is a good time to pause, and go back and read the Study tips from all the previous chapters. Re-learn anything which you have forgotten before you move on to Chapter 13.

Review and development

Practice questions for revision and exam preparation

Practice questions (13)

1. Sketch the function $y = x^3 - x^2$ in the range $-1 < x < 2$. Answer the following questions by observation and then use differentiation to confirm your results:
   (i) What is the range of values of x for which the gradient is negative?
   (ii) Where is the gradient zero?

2. Sketch the curve $y = x - 1/x$ in the range $-3 < x < 3$.
   (i) Show that the gradient of this curve is positive everywhere.
   (ii) When x gets very large, the gradient approaches a limit. Show what this limit is.

3. $y = (x - 2)^2 + 3$
   (i) Without using differentiation, write down the minimum value of this function.
   (ii) Write down the value of x at the minimum point.
   (iii) What is the value of y when $x = 0$?

(iv) Sketch the curve.

(v) Use symmetry to find the x–co-ordinate of the second point on the curve which has this same y-value.

(vi) Find the area bounded by the curve and the line $y = 7$.

4. A curve with gradient function $dy/dx = 4x - 3$ cuts the x-axis where $x = 1$.

(i) Find the equation of the curve.

(ii) Find the value of x where the curve cuts the x-axis again.

(iii) Find the co-ordinates of the stationary point on the curve and prove that this point is a minimum.

(iv) Sketch the curve.

(v) Estimate and then calculate the area between the curve and the x-axis. Explain why the result is negative.

5. The height, in metres above the ground, of a moving object is given by $y = -t^2 + 10t + 24$, where t is the time in seconds since the start of its journey.

(i) What is the height of the object at the start of its journey?

(ii) After how many seconds does it reach the ground?

(iii) What is its maximum height?

(iv) When does it reach its maximum height?

(v) Sketch the graph of the function.

6. $y = t^3$ and $t = x^2 - 1$.

(i) Find dy/dt and dt/dx.

(ii) Find $dy/dt \times dt/dx$ and express the result in terms of x.

(iii) Express y in terms of x.

(iv) Find dy/dx.

(v) Which rule of differentiation is being illustrated here?

7. The curve $y = x^3 - 2x^2 - x + 2$ cuts the x-axis in three places.

(i) Factorise the equation $x^3 - 2x^2 - x + 2 = 0$ and hence find the values of x at these three points.

(ii) Find the value of y when $x = 0$.

(iii) Sketch the curve.

(iv) Find the two values of x for which the gradient of the curve is -2.

(v) Find the equations of the tangent and the normal to the curve at the point $(2,0)$.

(vi) Find the area of the triangle formed by the y-axis and the tangent and the normal in (v).

Note: Integration is **not** required in part (vi).

8. Find the area between the curve in question 7 and the x-axis:

(i) in the range $0 < x < 1$

(ii) in the range $1 < x < 2$

Using these results, and your sketch from question 7, explain why integrating the function in the range $0 < x < 2$ would give an incorrect result, and state what you will do to avoid incorrect results in similar problems.

9. (i) Find the co-ordinates of the stationary points on the curve $y = x^3 - 6x^2 + 5$.

(ii) Find d^2y/dx^2 and use this to determine which is a maximum point and which is a minimum point.

(iii) Use trial and improvement to find one value of x for which $y = 0$.

(iv) Use your result from (iii) to factorise the equation $x^3 - 6x^2 + 5 = 0$ and hence write down all three values of x where the curve cuts the x-axis.

(v) Sketch the curve.

10. (i) Find the co-ordinates of the stationary points on the curve $y = 2x^4 - x^2$.

(ii) Use your answers to (i) to sketch the curve.

(iii) Modify the following statement to make it completely correct:

"The number of stationary points on the curve of an algebraic function is one less than the highest power of x in the function."

11. A prism whose cross section is an equilateral triangle of side a has length m and volume 250 cm³. Find the

values of a and m which will minimise the surface area of the prism.

12. (i) Sketch the curve $y = x^2$ in the range $0 < x < 3$.

   (ii) Express x as a function of y, and find the area bounded by the curve, the y-axis and the line $y = 4$.

   (iii) Find the value of h such that the area bounded by the curve, the y-axis and line $y = h$ is 18.

13. The equation of a circle with centre at $(0,0)$ and radius r is $x^2 + y^2 = r^2$.

   (a) Find dy/dx (i) in terms of y and x (ii) in terms of x only.

   (b) Find the co-ordinates of the two points on the circle with radius 5, where the gradient $= -0.5$.

14. The following observed values are produced during an experiment:

x	0.5	2.0	3.0	3.5	6.0
y	0	3	10	15	55

   (i) Plot these results on a graph.

   (ii) The experimenter believes that there is a quadratic relationship between x and y. This would mean that the data would fit an equation of the form $y = ax^2 + bx + c$. Assuming the experimenter is correct, use the first three pairs of values in the table to find the values of a, b and c.

   (iii) Check whether the remaining two pairs of values satisfy the equation with these values of a, b and c.

   (iv) Assuming the only values of x and y which are possible are those which satisfy this equation, find the minimum possible value of y.

15. (a) A curve is represented by the equation $y^2 = 2x^2 - 4x + 11$.

   (i) Find dy/dx in terms of x and y.

   (ii) Find the co-ordinates of the curve's only stationary point.

(iii) Find the two values of x for which $y^2 = 11$.

(iv) Using your answers to (ii) and (iii), sketch the curve.

(b) Now consider the general curve $y^2 = 2x^2 - 4x + k$.

(i) Find the value of k for which this curve has a minimum at the point $(1,2)$

(ii) What happens to the curve as k decreases and approaches the value 2?

(iii) Explain what happens to the function if $k = 2$.

16. An object is thrown with starting velocity 4 metres per second and comes to rest exactly 2 seconds later. The velocity is given by $v = 4 - t^2$.

(i) Sketch the curve for values of t in the range 0 to 2.

(ii) What is the velocity of the object 1 second after being thrown?

(iii) How far would the object travel in 2 seconds if it maintained its initial velocity throughout?

(iv) How far would the object travel in 2 seconds at an average speed of 2 metres per second?

(v) From your sketch, estimate the average speed of the object.

(vi) Using your answers to these questions, give a rough estimate of the distance travelled by the object in 2 seconds.

(vii) Find the exact distance travelled by integrating the function $v = 4 - t^2$ from $t = 0$ to $t = 2$.

17. A car is travelling at 100 metres per second (this is just under 17 miles per hour).

The driver puts her foot on the brake and the speed of the car from this moment is given by:

$$v = 100 - 4t^2$$

(i) For how many seconds does the car continue to move before coming to a stop?

(ii) How far does the car travel from the moment the brakes are applied?

(iii) New legal specifications for brakes are introduced, stating that a car travelling at 100 m/s must be able to stop in less than 3 seconds. Your answer to (i) shows that this car does not meet this requirement. If the speed of a car with legal brakes is given by $v = 100 - kt^2$, then in order for the speed to be reduced more quickly, k must be greater than 4. Find a whole number value of k for which the car will stop in less than 3 seconds.

(iv) Using this value of k, find by integration the distance the car travels during the 3 seconds.

18. Investigate the function $y = (x - 2)(x - 5)$:

   (i) At which two points does the curve cut the x-axis?

   (ii) At which point does the curve cut the y-axis?

   (iii) Find the stationary point and state whether this is a maximum or a minimum point.

   (iv) Sketch the curve.

   (v) Estimate, then calculate, the area between the curve, the x-axis, and the y-axis.

   (vi) Estimate, then calculate, the area between the curve and the x-axis in the range $2 < x < 5$.

   (vii) Explain why your answer to (vi) is negative.

19. A window cleaner operates on a large housing estate. He has worked out that his costs (travel and materials) average out at £2 per house. He finds that if he charges £10 per house he gets 20 customers in a week, but if he charges £5 per hour he gets 60 customers in a week.

   (i) Assuming that there is a linear relationship between p, the price he charges, and n, the number of customers he gets, find an equation connecting n and p.

   (ii) Find an expression for P, the amount of profit he makes in a week, in terms of n and p.

(iii) Use your equation from (i) to express P in terms of p only.

(iv) Use differentiation to find:

the price the window cleaner should charge so that his profit is a maximum,

the number of customers he will get at this price,

the profit he will make at this price.

(v) How much would you advise the window cleaner to charge in practice, and why?

20. (i) Find the gradient function of the curve $y = -x^3 + 6x + 4$.

(ii) Find the gradient of the curve at the point where it cuts the y-axis.

(iii) Choose a small positive value of x, and a small negative value of x, and use these values to find the gradient of the curve near the point $(0,4)$.

(iv) Find d^2y/dx^2 and its value at the point $(0,4)$.

(v) Describe in words what happens to the gradient at the point $(0,4)$.

(vi) Sketch, on separate graphs, the curve and its gradient function.

Appendix – Revision topics

Note: In this Chapter the answers to the exercises are given at the end of each topic.

1. Speed, distance and time

In this section we give a summary of this topic, showing what you need to know in order to move on to differentiation. The essential point to understand is that we don't need calculus to work with constant, or average speeds. These can be easily explored using simple arithmetic and straight line graphs. But when we are looking at varying speeds (which in real life is nearly always!) we need to use calculus.

The three quantities speed, distance and time are interdependent. This means that if you know two of them, you can calculate the third one. To do this, we use this relationship:

$$\text{distance} = \text{speed} \times \text{time}$$

which can also be written:

$$\text{speed} = \text{distance} \div \text{time}$$

or

$$\text{time} = \text{distance} \div \text{speed.}$$

Constant (average) speed

The relationship between speed, distance and time, which we are using here, is based on an assumption of constant speed. In reality, speed is rarely constant. Whether driving, walking, running, or cycling, etc, we normally slow down and speed up during a journey, even a very short one. But that doesn't matter if we want results which are based on an average speed. And for mathematical purposes, when we assume an average speed, this is the same as a constant speed.

Example 14.1.1

You go for a 4-mile walk and it takes you 2 hours.

Known quantities:

Distance = 4 miles   Time = 2 hours

Calculate:

Speed = 2 miles per hour

What does "2 miles per hour" mean? When you go for a walk, it's very unlikely that you would walk at exactly the same speed for 2 hours. You might stop and admire the view now and again! You might walk faster at the beginning and more slowly towards the end. BUT you have walked at an **average** speed of 2 miles per hour.

So, as long as we are working with average, or constant speeds, we can use the relationship between speed, distance and time, to calculate whichever of these quantities is unknown.

Exercise 14.1.1

1. How long would a runner take to run 1000 metres at a speed of 8.5 metres per second?

2. You set off from home at 9.00 a.m. to drive to a city which is 250 miles away. You drive at an average speed of 60 mph. Will you get there by 1.00 p.m.?

3. Tom walks to work. His office is 1½ miles away from his house. He can walk at a steady speed of 3 miles per hour. He has to be at work at 9.00 a.m.

   a) At what time should Tom leave his house?

   b) If Tom leaves the house 5 minutes late and walks at his normal speed for 20 minutes, how fast will he have to walk (or run) to complete the journey and arrive on time?

Graphs

Journeys can be easily represented on a graph.

On a distance/time graph, the horizontal axis is used for time, and the vertical axis is used for distance.

The slope (gradient) of the graph shows the speed.
Here is an example, showing a journey of 3 km which takes
1 hour. The average speed for the journey is given by the
gradient of the line, which is 3 km ÷ 1 hr = 3 km/hour.

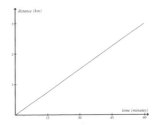

Fig 14.1.1

Exercise 14.1.2

1. Plot a graph for Tom's journey to work, as in question
   3(a) above.

   The horizontal axis should show minutes from 0 to 30.
   The vertical axis should show distance up to 1½ miles.
   In the normal situation, where Tom leaves his house
   on time, the graph will be a straight line from (0,0) to
   (30,1½). The gradient of the line is:

   $$\frac{\text{height}}{\text{width}} = \frac{1\frac{1}{2}\text{ miles}}{30\text{ mins}} = \frac{1\frac{1}{2}\text{ miles}}{\frac{1}{2}\text{ hour}} = 3\text{ mph}$$

   [Note how the units work. Dividing distance (measured
   in miles) by time (measured in hours) gives speed
   measured in miles per hour.]

2. Plot a graph for Tom's journey when he leaves the
   house 5 minutes late.

   The walking part of the journey is a straight line from
   (5, 0) to (25, 1).

   The running part of the journey is a straight line from
   (25,1) to (30,1½).

   Notice how the second part of the journey is
   represented by a line with a steeper gradient.

Find the speed of each part of the journey by working out the gradients.

(For revision of gradients, see the next section of this Appendix).

Answers to exercise 14.1.1

1. Use the formula:

$$\text{time} = \text{distance} \div \text{speed}$$

This gives:

$$\text{time} = 1000 \div 8.5 = 117.65 \text{ seconds}$$
$$= 1 \text{ minute } 57.65 \text{ seconds}$$

2. Use the formula:

$$\text{distance} = \text{speed} \times \text{time}$$

This gives:

$$\text{distance} = 60 \text{ mph} \times 4 \text{ (hours)} = 240 \text{ miles.}$$

No, you won't get there by 1.00 p.m.! You will be 10 miles away, and if you continue at the same speed you will need a further 10 minutes to get there.

[Note that 60 mph is the same as 1 mile per minute, since there are 60 minutes in one hour.]

3. a) time = distance ÷ speed = 1½ ÷ 3 = ½ hour.

So Tom should leave his house at 8.30. a.m.

b) We are working in hours not minutes, so first convert 20 minutes to 1/3 of an hour.

distance = speed × time = 3 × 1/3 = 1 mile.

So Tom has only 5 minutes left in which to cover the remaining ½ mile.

Convert 5 minutes to 1/12 of an hour.

speed = distance ÷ time = ½ ÷ (1/12) = ½ × 12 = 6 miles per hour.

This is certainly a running, not a walking speed!

Answers to exercise 14.1.2

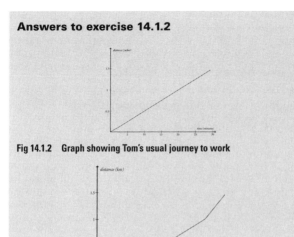

Fig 14.1.2   Graph showing Tom's usual journey to work

Fig 14.1.3   Graph showing Tom's journey when he leaves 5 minutes late

2. Straight lines and gradients

A straight line can be represented on an x,y graph by an equation in the form $y = mx + c$.

In this section we revise straight line graphs and the relationship between the equation of the line, and its gradient.

Example 14.2.1

The equation $y = x$ passes through the points $(-1,-1)$, $(0,0)$, $(1,1)$ etc.

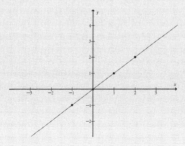

Fig 14.2.1

To calculate the gradient of a straight line, we work as follows:

Choose a suitable section of the line.
Find the vertical height of this section (h).
Find the horizontal distance of this section (d).
Then the gradient $= h \div d$.

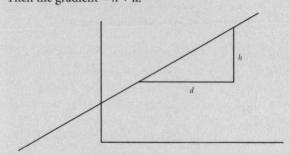

Fig 14.2.2

The result will be the same for any section of the same straight line.

For the line $y = x$, between the points (0,0) and (1,1), h and d are both 1.

So the gradient of the line $y = x$ is 1.

Exercise 14.2.1

1. On the same graph, plot the two lines $y = x + 1$ and $y = x - 2$.
   a) What are the gradients of these lines?
   b) At which point does each line cut the y-axis?
2. On a new graph, plot the lines $y = -x$ and $y = 3 - x$.
   a) What are the gradients of these lines?
   b) At which point does each line cut the y-axis?
3. Compare the equations of the lines in questions 1 and 2 with the general equation of the straight line:

$$y = mx + c$$

   Which properties of the line do m and c correspond to?
   (Note that in question 2, the equation $y = 3 - x$ could alternatively be written $y = -x + 3$.)

4. Sketch these lines:

a) $y = -2x + 2$
b) $y = \frac{1}{2}x - 1$

["Sketch" means that you do not need to calculate co-ordinates or plot points.
Use your knowledge of the properties of straight lines to draw a sketch.
You do need to draw the axes, of course, but graph paper isn't necessary for a sketch.
You need sufficient indication of the scale on each axis but you don't need to label the axes with numbers all the way along.]

Equations in different forms

The equation of a straight line has 3 terms:

a term in x
a term in y
a constant term (which may be zero).

If the equation is in the form $y = mx + c$, then we can immediately see the gradient and the y-intercept.

If the equation has the terms in a different arrangement, then we need to re-arrange it before we can see its gradient and y-intercept.

Example 14.2.2
$2y + x = 4$ is the same equation as $y = -\frac{1}{2}x + 2$
so the gradient of this line is $-\frac{1}{2}$ and the y-intercept is 2.

Exercise 14.2.2
Find the gradients and the y-intercepts of each of these lines:

a) $3y - 2x = 6$ b) $x + y = 4$
c) $2x - 4y = 3$ d) $3x + 6y = 2$

Horizontal and vertical lines
The gradient of a horizontal line is zero. So the equation of a horizontal line has $m = 0$.

The horizontal line with y-intercept 1 has equation $y = 1$. All equations of the form $y = c$ represent horizontal lines (i.e. parallel to the x-axis).

The gradient of a vertical line is infinite (undefined). So the equation of a vertical line has no y-term. The vertical line which cuts the x-axis at $x = 2$ has equation $x = 2$

All equations of the form $x = a$ represent vertical lines (i.e. parallel to the y-axis).

Answers to exercise 14.2.1

1. a) Both lines have gradient 1.
   b) $(0,1)$, $(0,-2)$
2. a) Both lines have gradient -1.
   b) $(0,0)$, $(0,3)$
3. m is the gradient. c is the y-coordinate of the point where the line cuts the y-axis. This is called the y-intercept.
4. a)

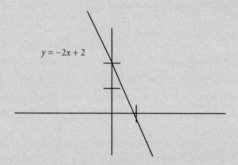

$y = -2x + 2$

Fig 14.2.3

   b)

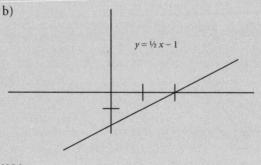

$y = \frac{1}{2}x - 1$

Fig 14.2.4

Answers to exercise 14.2.2

a) $3y - 2x = 6$

   $3y = 2x + 6$

   $y = \dfrac{2x}{3} + 2$ gradient = 2/3, y-intercept = 2

b) $x + y = 4$

   $y = -x + 4$ gradient = –1, y-intercept = 4

c) $2x - 4y = 3$

   $4y = 2x - 3$

   $y = \frac{1}{2}x - \frac{3}{4}$ gradient = ½, y-intercept = – ¾

d) $3x + 6y = 2$

   $6y = -3x + 2$

   $y = -\frac{1}{2}x + 1/3$ gradient = – ½, y-intercept = 1/3

3. Finding points of intersection

A straight line (or curve) can be thought of as an infinite set of points. Every point on the straight line (or curve) has co-ordinates (x,y) which fit its equation. The point where two lines intersect has co-ordinates which fit the equations of both of the lines. We can find the co-ordinates of the point of intersection by drawing the two lines or curves on the same graph, or by solving the equations using algebra.

A curve and a line may intersect or only touch (if the line is a tangent). There may be no meeting point at all, in which case the equations we create will have no solution.

In this section, we investigate these different cases using both graphical and algebraic solutions.

Example 14.3.1

Find the point of intersection of the lines $y = 3x - 1$ and $y = x + 2$.

We can do this graphically by plotting both lines on the same graph:

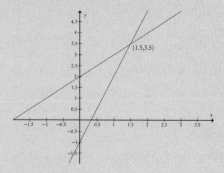

Fig 14.3.1

By choosing a suitably small scale, we can see that the lines intersect at the point (1.5,3.5).

Alternatively, we can find this point by solving the two equations simultaneously:

$$y = 3x - 1$$
$$y = x + 2$$

Subtraction gives $0 = 2x - 3$; so $x = 1.5$, and substituting back gives $y = 3.5$

The algebraic method is often quicker than the graphical method.

Exercise 14.3.1

1. Find the point of intersection of the lines $2y = 4x + 1$ and $x = y = 2$
   (i) by the graphical method
   (ii) by algebra.
2. Write down the equations of any two lines which have no point of intersection.
3. Write down the equations of any two lines which intersect at the origin.

Example 14.3.2

Find the co-ordinates of the points where the line $y = 5x + 8$ cuts the curve $y = x^2 - 2x$.

As we cannot guarantee an accurate answer by the graphical method where a curve is involved, we will use algebra. At the points of intersection, the x and y co-ordinates are equal, so we start by equating the two expressions for y:

$$x^2 - 2x = 5x + 8$$

and this simplifies to the quadration equation:

$$x^2 - 7x - 8 = 0$$

Factorising,

$$(x - 8)(x + 1) = 0$$

giving

$$x = 8 \text{ and } x = -1$$

Substituting these x-values into the equation $y = 5x + 8$ gives the y-values 48 and 3 respectively.
(Substituting into $y = x^2 - 2x$ will give the same answers but the straight line equation is easier to use.)
So the points of intersection are (8,48) and (–1,3).

Example 14.3.3

Find the co-ordinates of the points where the line $y = -4x - 4$ cuts the curve $y = x^2$.
Using the algebraic method again, we have $x^2 = -4x - 4$ which gives the quadratic equation

$$x^2 + 4x + 4 = 0$$

Factorising,

$$(x + 2)(x + 2) = 0$$

giving

$$x = -2$$

In this example there is one solution only. This indicates that the line $y = -4x - 4$ touches but does not intersect the curve $y = x^2$. The line is a tangent to the curve and the point of contact is (–2,4).

Exercise 14.3.2

1. The line $y = -4x - 6$ has no point of contact with the curve $y = x^2$.

   Write down the quadratic equation which we would use to try to find a point of contact and explain what happens when you try to solve this equation.

2. Sketch the line $y = 5x + 8$ and the curve $y = x^2 - 2x$ on the same graph and check that the points of intersection are as in Example 14.3.2.

3. Without drawing a graph, investigate whether the line $y = 2x + 5$ and the curve $y = 4 - x^2$ have two points of intersection, one point of contact, or no points of intersection.

Answers to exercise 14.3.1

1. (i)

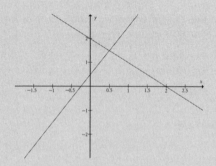

Fig 14.3.2

(ii) $2y = 4x + 1$

$x + y = 2$

Doubling the second equation and rearranging it gives

$$2y = -2x + 4$$

and now we can eliminate y by writing

$$4x + 1 = -2x + 4$$

The solution of this equation is $x = 0.5$ and substitution into $x + y = 2$ gives $y = 1.5$

2. You can choose any pair of parallel lines.
   Parallel lines have the same gradient, so your answer should be two equations in the form

$$y = mx + c_1$$

and

$$y = mx + c_2$$

where $c_1 \neq c_2$

3. You can choose any two equations in the form

$$y = m_1 x$$

and

$$y = m_2 x$$

where $m_1 \neq m_2$

Answers to exercise 14.3.2

1. $x^2 + 4x + 6 = 0$
   This equation does not factorise. Applying the formula, we see that "$b^2 - 4ac$" is negative ($= -8$).
   Therefore there are no real solutions to this equation. This corresponds to there being no point of contact between the curve $y = x^2$ and the line $y = -4x - 6$.

2.

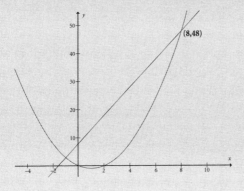

(8,48)

Fig 14.3.3

3. The equation to be solved is $4 - x^2 = 2x + 5$, which reduces to $x^2 + 2x + 1 = 0$.

This equation has only one solution, $x = -1$. The line and the curve have one point of contact.

4. Practising algebra

Algebra is a huge topic. Here we restrict ourselves to explaining the techniques you need in order to work through this book: multiplying out brackets, factorising quadratic expressions, and algebraic fractions.

Multiplying out brackets

In mathematics we often need to multiply together two or more expressions in brackets.

Example 14.4.1

First, we'll look at a numerical example. We want to simplify $(7 + 2)(4 + 3)$ to one number.

Because we have only numbers here (no unknowns), we would normally follow the rule which says "do what's in the brackets first", and this gives $9 \times 7 = 63$.

But it would also be possible to do the multiplications first and the adding afterwards. We do this by multiplying each term in the first bracket by each term in the second bracket, and adding up all the results.

The multiplications are: 7×4 7×3 2×4 2×3
and the answers to these are: 28   21   8   6
Adding these up gives   63.

Well, nobody in their right mind would do a sum like this by such a long method!

But if some of the numbers are replaced by unknowns, it is not possible to "do what's in the bracket first".

Example 14.4.2

We want to simplify $(x + 5)(x + 2)$. Clearly this can't reduce to one number, but we want to get rid of the brackets and combine any terms which can be added (or subtracted).

Following the method shown in Example 14.4.1, we have:

Multiplications	$x \times x$	$x \times 2$	$5 \times x$	5×2
Answers	x^2	$2x$	$5x$	10
Adding up	$x^2 + 7x + 10$			

Example 14.4.3

This one has two minus signs. Remember that every sign (+ or −) belongs to the term which follows it.
Simplify $(x - 5)(y - 3)$:

Multiplications	$x \times y$	$x \times -3$	$-5 \times y$	-5×-3
Answers	xy	$-3x$	$-5y$	15
Adding up	$xy - 3x - 5y + 15$			

Exercise 14.4.1

Multiply out the brackets and simplify the resulting expression where possible.
(a) $(x + 2)(x + 1)$ (b) $(a - 3)(b + 1)$ (c) $(c - d)(c - 2d)$
(d) $(p - q)(p + q)$ (e) $(2f + g)(g - f)$ (f) $(x^2 + 1)(x - 1)$

Factorising quadratic expressions

This is the reverse of multiplying out brackets. This technique is needed for solving quadratic equations.
We start with a quadratic expression and express it as a product of two linear expressions.

Example 14.4.4

a) Factorise $x^2 + 5x + 6$.

Step 1: Decide on the signs.
In this case, both signs are positive so the factors will be in the form $(x + a)(x + b)$
Step 2: Find two numbers which multiply together to make 6. The possibilities are 1×6 or 2×3.
Step 3: Choose the pair which add to 5, which is 2 and 3.
Step 4: Write the expression in its factorised form and multiply to check:

$$(x + 2)(x + 3)$$

Multiplications	$x \times x$	$x \times 3$	$2 \times x$	2×3
Answers	x^2	$3x$	$2x$	6
Adding up	$x^2 + 3x + 2x + 6$			

and this is correct.

b) Factorise $x^2 - 5x + 6$.

This is nearly the same problem. Compare this process with example a):

Step 1: Decide on the signs.

In this case, the middle term is negative and the last term is positive, so the factors will be in the form $(x - a)(x - b)$

Step 2: Find two numbers which multiply together to make 6. The possibilities are 1×6 or 2×3.

Step 3: Choose the pair which add to 5, which is 2 and 3.

Step 4: Write the expression in its factorised form and multiply to check:

$$(x - 2)(x - 3)$$

Multiplications	$x \times x$	$x \times -3$	$-2 \times x$	-2×-3
Answers	x^2	$-3x$	$-2x$	$+6$
Adding up	$x^2 - 3x - 2x + 6$			

and this is correct.

c) Factorise $x^2 - 5x - 6$.

Slightly different this time.

Step 1: Decide on the signs.

In this case, both the middle and the last term are negative.

The factors must be in the form $(x + a)(x - b)$ to make the last term negative.

Step 2: Find two numbers which multiply together to make −6.

The possibilities are -1×6 or 1×-6 or -2×3 or 2×-3.

Step 3: Choose the pair which add to –5, which is 1 and –6.

Step 4: Write the expression in its factorised form and multiply to check:

$$(x - 6)(x + 1)$$

Exercise 14.4.2

1. a) Complete Example 14.4.4 c); multiply out the brackets $(x - 6)(x + 1)$ and check the result.

   b) Factorise $x^2 + 5x - 6$; use Example 14.4.4

   c) for comparison.

2. Factorise these quadratic expressions. If, at Step 3, you find that none of your pairs of numbers adds up to the coefficient of x, then either you have missed a pair of numbers, or the quadratic expression has no factors.

   a) $x^2 + 2x + 1$ b) $x^2 + 3x + 2$

   c) $x^2 + 17x + 70$ d) $x^2 + 8x + 12$

   e) $x^2 - 2x + 1$ f) $x^2 - 4x + 3$

   g) $x^2 - 11x + 24$ h) $x^2 - 13x + 44$

   i) $x^2 - 3x - 4$ j) $x^2 + 3x - 4$

   k) $x^2 - 2x - 8$ l) $x^2 + 2x - 9$

3. a) Factorise $x^2 - 1$. (Write the expression as $x^2 + 0x - 1$ and proceed as before).

   b) Factorise $x^2 - 4$ without doing any working.

Algebraic fractions

We can reduce numerical fractions to their simplest terms by dividing both the numerator and the denominator by a common factor. For example:

$$\frac{8}{12} = \frac{2}{3}$$

If the numerator is a series of terms to be added or subtracted, we can still simplify the fraction by dividing each term in the numerator by the same factor as we divide the denominator. For example:

$$\frac{12+24-16+20}{8} = \frac{3+6-4+5}{2}$$

Both the sum on the left and the sum on the right give the result 10. The sum on the right has every term dvided by 4 and becomes a simpler calculation. [In practice, there would be no reason for leaving the numerator as separate terms. But this example shows that we can simplify the fraction whether it has one or several terms in the numerator.]

Now some algebraic examples:

$$\frac{ab}{bc} = \frac{a}{c}$$

(Numerator and denominator divided by b)

$$\frac{a^2d + ad + abc}{bd} = \frac{a^2 + a + acd}{b}$$

(Numerator and denominator divided by d)

$$\frac{(x+1)^3}{(x+1)^2} = x+1$$

(Numerator and denominator divided by $(x + 1)$)

Exercise 14.4.3

Simplify these fractions.
In d) you need to multiply the bracket out first.
In e) you need to factorise the numerator first.

a) $\dfrac{25a^2}{5a}$

b) $\dfrac{x^3 - x^2 + x}{xy}$

c) $\dfrac{4a^3 - 3a}{2a}$

d) $\dfrac{(x+1)(x-3)+3}{x}$

---

e) $\dfrac{x^2-1}{x-1}$

f) $\dfrac{ab+bc-ab^2}{ab}$

g) $\dfrac{(x+1)(x+2)-(x+1)}{(x+1)}$

h) $\dfrac{x^2-5x+6}{(x-2)}$

Answers to exercise 14.4.1

(a) x^2+3x+2 (b) $ab+a-3b-3$ (c) $c^2-3cd+2d^2$

(d) p^2-q^2 (e) $fg-2f^2+g^2$ (f) x^3-x^2+x-1

Answers to exercise 14.4.2

1. a) x^2-5x-6 b) $(x+6)(x-1)$
2. a) $(x+1)(x+1)$ or $(x+1)^2$ b) $(x+1)(x+2)$
   c) $(x+7)(x+10)$ d) $(x+6)(x+2)$
   e) $(x-1)^2$ f) $(x-1)(x-3)$
   g) $(x-3)(x-8)$ h) No factors
   i) $(x+1)(x-4)$ j) $(x-1)(x+4)$
   k) $(x-4)(x+2)$ l) No factors
3. a) $(x-1)(x+1)$ b) $(x-2)(x+2)$

Answers to exercise 14.4.3

a) $5a$ b) $\dfrac{x^2-x+1}{y}$ c) $2a-\dfrac{3}{2}$ d) $x-2$

e) $x+1$ f) $1+c/b-b$ g) $(x+2)-1=x+1$ h) $x-3$

5. Understanding powers and surds

In this section we deal with powers (also called indices), which may be positive, negative, or fractions.

We also explain surds, which are a particular application of fractional powers.

Repeated multiplication

A power indicates that a number is to be multiplied by itself.

If the power is 2, that means we write the number down twice and put a × sign in between.

If the power is 3, that means we write the number down three times and put × signs in between.

So x^2 **is the same as** $x \times x$ and x^3 **is the same as** $x \times x \times x$.

This simple rule is only about saving time and space – it's a form of shorthand.

But unfortunately it can lead to errors if you are not careful. For example, $3^2 = 3 \times 3 = 9$ but this isn't always the answer which appears when students are under pressure in exams!

> **Exercise 14.5.1**
>
> Some practice in using powers. Give a numerical answer for each of these:
>
> a) 2^3 b) 3^3 c) 10^2
>
> d) 5^4 e) 0.5^2 f) 0.1^3
>
> g) 6^1 (write the number down once!)
>
> h) 1^5

Zero and negative integer powers

The rule which says that the power is the "number of times we write the number down" is not only a bit long-winded, but it only works if the power is a positive integer.

So what does x^0 mean? And what does x^{-1} mean?

To answer these questions, we write down a table of powers in descending order, and extend it to zero and negative integers. We can use any base number for this purpose, so let's choose 2. Starting at 2^4 and working downwards, we have:

$$2^4 = 16$$
$$2^3 = 8$$
$$2^2 = 4$$
$$2^1 = 2$$

Each time we write the next line in this table, the power decreases by 1 and the value on the right is halved.

So the next line must be:

$$2^0 = 1$$

and then

$$2^{-1} = \tfrac{1}{2}$$
$$2^{-2} = \tfrac{1}{4}$$

and so on.

We could have chosen any base number x in place of 2, and we would always get $x^0 = 1$.

And for negative integer powers, we see from the table that $2^{-1} = 1/2$, $2^{-2} = 1/2^2$ and we can generalise this observation to $2^{-n} = 1/2^n$.

Exercise 14.5.2
Find the values of:
a) 3^{-1} b) 10^0 c) 4^{-2} d) 0.5^{-1}
e) 10^{-4} f) 0.1^{-2} g) 2^{-4} h) 1^{-n}

Fractional powers

We also need to understand powers such as ½, ⅓, ⅔, ¾ etc.

First we need to recall the "add powers" rule. This operates when we have the same base number, let's say 2, raised to different powers and we want to multiply. For example, $2^3 \times 2^4$. Written out in full, this sum is:

$$(2 \times 2 \times 2) \times (2 \times 2 \times 2 \times 2)$$

[brackets only shown for clarity]

which is 2^7. So $2^3 \times 2^4 = 2^7$. We can generalise this to the rule $x^a \times x^b = x^{a+b}$.

Now let's look at the case where a and b are both ½. Then we have $x^{1/2} \times x^{1/2} = x^{1/2 + 1/2} = x^1 = x$.

So $x^{1/2} \times x^{1/2} = x$ must be true for any value of x.

This statement says that a number (x) can be written as a number ($x^{1/2}$) multiplied by itself.

The only possibility is that $x^{1/2}$ is the square root of x. For example, if $x = 9$, then $x^{1/2} = 3$.

If $x = 49$, then $x^{1/2} = 7$.   If $x = 2$, then $x^{1/2} = \sqrt{2} = 1.414$ (to 3 d.p.).

We can use a similar argument to show that $x^{1/3}$ represents the cube root of x, and in general, $x^{1/n}$ represents the nth root of x.

> **Exercise 14.5.3**
> Write down the values of:
> a) $4^{1/2}$ b) $100^{1/2}$ c) $3^{1/2}$ (correct to 3 d.p.)
> d) $8^{1/3}$ e) $16^{1/4}$ f) $81^{1/4}$

Combined powers

The "power to a power" rule $(x^a)^b = x^{ab}$ can be used to find the value of an expression such as $(2^3)^2$.

$(2^3)^2 = 2^3 \times 2^3 = 2^6$ (using the "add powers" rule), and $2^6 = 64$.

We can use this rule if a or b (or both) is a fraction. For example, $(16^{1/2})^3 = 16^{3/2}$.

$16^{1/2} = 4$, $4^3 = 64$. So $16^{3/2} = 64$. When we apply this rule to fractional powers we usually start with a power such as 3/2 and have to split it into factors $(3 \times \frac{1}{2})$. So if we need to evaluate $4^{2/3}$ we write this as $(4^2)^{1/3} = 16^{1/3} = \sqrt[3]{16} = 2.520$ to 3 d.p.

Surds

Surds are roots of numbers (square roots, cube roots etc) which cannot be written as exact numbers because they have infinite non-repeating decimal places. For example, $\sqrt{2} = 1.41421356$. It is not practical to write the square root of 2 in this form, so we simply write it as $\sqrt{2}$. Similarly, $\sqrt{3}$, $\sqrt{5}$, and many more square roots do not have rational values, so we can work with them in this form until we come to the end of a problem when we would probably want to round off the answer to a given degree of accuracy.

We would not leave $\sqrt{4}$ in this form, because $\sqrt{4}$ is an exact number, 2. The same is true for any number which is a perfect square. Some surds can be factorised and simplified, for example $\sqrt{8} = \sqrt{(4 \times 2)} = 2\sqrt{2}$. Practise in manipulation of surds is not required for the work in this book, but if you want to add this to your repertoire of mathematical skills, you will find some help with it in one of the websites listed at the end of this book.

An important point to remember is that the square root sign $\sqrt{}$ and the power $\frac{1}{2}$ are interchangeable:

$$\sqrt{x} = x^{1/2}$$

and $\sqrt{x}$ needs to be written as $x^{\frac{1}{2}}$ in order to differentiate it.

Exercise 14.5.4
Find the value of:
a) $16^{3/4}$ b) $25^{3/2}$ c) $1^{3/2}$
d) $1000^{2/3}$ e) $8^{1/2}$ (correct to 2.d.p) f) $4^{5/2}$

Answers to exercise 14.5.1
a) 8 b) 27 c) 100 d) 625
e) 0.25 f) 0.001 g) 6 h) 1

Answers to exercise 14.5.2
a) ⅓ b) 1 c) 1/16 d) 2
e) 0.0001 f) 100 g) 1/16 h) 1

Answers to exercise 14.5.3
a) 2 b) 10 c) 1.732 d) 2
e) 2 f) 3

Answers to exercise 14.5.4
a) 8 b) 125 c) 1 d) 100
e) 2.83 f) 32

6. Basic co-ordinate geometry

Co-ordinate Geometry is about working with lines and curves on an (x,y) graph.
The techniques we show here are:
Finding the equation of a straight line.
Using the relationship between the gradients of perpendicular lines

The equation of a straight line

Two points are enough to define a straight line. First we can use the co-ordinates of these two points to calculate the gradient, m. Then we can use this gradient value and one of the pairs of co-ordinates to find the y-intercept, c. (This is not the only method, but it has the advantage that the only formula you need is the one you already know, $y = mx + c$.)

Example 14.6.1

We want to find the equation of the line joining the two points $(-2,3)$ and $(1,4)$.

Step 1: Calculate the gradient:

$$m = (4-3)/(1-(-2)) = 1/3$$

Step 2: Substitute $m = 1/3$ and the point $(1,4)$ into the equation $y = mx + c$:

$$4 = \tfrac{1}{3} + c$$

This gives $c = 3\tfrac{2}{3}$ and so the equation of the line is

$$y = \tfrac{1}{3}x + 3\tfrac{2}{3}.$$

Exercise 14.6.1

Find the equations of the lines joining these pairs of points. Check that your equation is correct in each case by substituting *both* pairs of co-ordinates.

a) $(1,1)$ and $(2,6)$ b) $(-2,-4)$ and $(0,4)$
c) $(-1,3)$ and $(2,5)$ d) $(0,-3)$ and $(1,-2)$

Alternative formulae

There are two formulae which can be used to find the equation of a straight line.

The first can be used when two points (x_1,y_1) and (x_2,y_2) are known:

$$\frac{y - y_1}{y_2 - y_1} = \frac{x - x_1}{x_2 - x_1}$$

The second can be used when one point (x_1,y_1) and the gradient m are known:

$$y - y_1 = m(x - x_1)$$

You can practise using either of these formulae by working again through Exercise 14.6.1.

Perpendicular gradients

In Section 2 of this chapter we saw how to calculate the gradient of a straight line (see Fig 14.2.2).

Now let's look at this again and include a second line at right angles to the first:

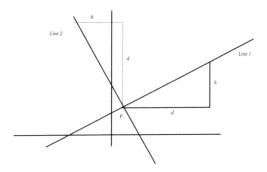

Fig 14.6.1

The gradient of *Line 1* is $h \div d$. If we rotate the triangle through 90° anticlockwise about the point P, we have a triangle which we can use to find the gradient of *Line 2*. This triangle has height is d and width h. But for the purposes of calculating the gradient of *Line 2*, we note that the width of the triangle is measured in the negative direction, i.e. to the left. So the gradient of *Line 2* is $d \div (-h) = -(d \div h)$.

Comparing the gradients of *Line 1* and *Line 2*, we see that if we multiply the gradients together we get −1:

$$\frac{h}{d} \times \frac{-d}{h} = -1.$$

[When two numbers multiply together to −1, we say that one is the **negative reciprocal** of the other.]

Example 14.6.2

What is the gradient of a *line* which is perpendicular to the line $y = 3x + 1$?

The negative reciprocal of 3 is −⅓. Any line with equation $y = -⅓ x + c$ will be perpendicular to the line $y = 3x + 1$.

More generally, all the lines $y = -\frac{1}{3}x + c$ (for any value of c) will be perpendicular to all the lines $y = 3x + c$ (for any value of c).

Exercise 14.6.2

Find the gradients of lines which are perpendicular to these lines:

a) $y = 2x + 3$ b) $y = x - 1$ c) $2y + x = 4$

d) $2x + y = 6$

Answers to exercise 14.6.1

a) $y = 5x - 4$ b) $y = 4x + 4$ c) $y = \frac{2}{3}x + 3\frac{2}{3}$

d) $y = x - 3$

Answers to exercise 14.6.2

a) $-\frac{1}{2}$ b) -1 c) 2 d) $\frac{1}{2}$

Answers

Practice questions (1.1)

1. $D = 0.8k$
2.

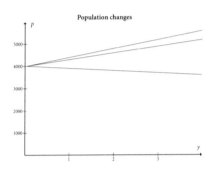

Fig A1.1

    (iv) $P = 4000 + 400Y$
         $P = 4000 - 100Y$
         $P = 4000 + 300Y$

Practice questions (1.2)

1. $\frac{dP}{dY} = 400$ $\frac{dP}{dY} = -100$ $\frac{dP}{dY} = 300$
2. $\frac{ds}{dt}$ This will represent the velocity (speed).
3. $C = 50 + 20n$
4. $C = 2\pi r$.

   $\frac{dC}{dr}$ represents the rate of increase of the circumference relative to the radius.

   For each unit increase in the radius, the circumference increases by 2π.
5. $\frac{dT}{dk}$ is the rate at which the cooking time changes for different weights of chicken.

   $\frac{dT}{dk} = 30$ minutes per kilo.
6. (i) Yes, m is the gradient and is equal to dy/dx.

   (ii) Yes, see (i)

(iii) No, the gradient is unaffected by the value of c. For any one value of m, there can be infinitely many different values of c in the equation $y = mx + c$.

Progress questions (1)

1. (i) $T = 25 + 20W$
    (ii) 145 minutes.
    (iii) $\frac{dT}{dW}$ is the rate of change of the cooking time for different weights of turkey.
2. (a)   (i) 4 seconds (ii) 2.8 seconds
   (b)   (i) $V = u - 15t$
         (ii) $\frac{dV}{dt} = -15$ which is the deceleration of the car in mph/second.

Practice questions (2.1)

1. The gradient of $y = x^2$ at the origin is 0.
2. The gradient is negative in the second quadrant.
3. The gradient is large and positive when x and y are large and positive. It is large and negative when x and y are large and negative.
4. No.

Practice questions (2.2)

The cubic curve

1. 0.
2. Nowhere.
3. The gradient is large and positive when x and y are large and positive. It is also large and positive when x and y are large and negative.
4. Yes. For every positive value of x, the curve has the same gradient at the corresponding negative value of x.

The rectangular hyperbola

1. This curve does not pass through the origin. When $x = 0$, y is infinite (positive or negative) and the gradient is infinitely steep and negative.
2. The gradient is negative at every point on the curve.
3. See answer to question 1.
4. See answer to question 4 for the cubic curve.

Progress questions (2)

1.

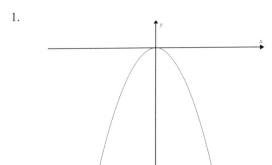

Fig A2.1

The equation of this curve is $y = -x^3$

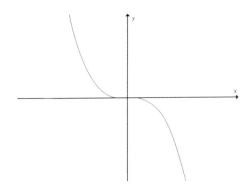

Fig A2.2

The equation of this curve is $y = -x^2$

Practical assignment (2)

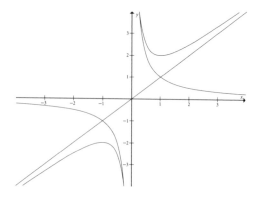

Fig A2.3

x	Gradient of $y = x$	Gradient of $y = 1/x$	Gradient of $y = x + 1/x$
1	1	−1	0
−1	1	−1	0

When functions are added, their gradients can be added too.

Note: This is only an exploration! We will not be providing proof of this property of functions.

Practice questions (3)

a)

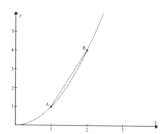

Fig A3.1

Gradient of $AB = 3$

b)

A		B				Gradient of AB
x	y	x	y	dx	dy	dy/dx
1	1	2	4	1	3	3
1	1	1.8	3.24	0.8	2.24	2.8
1	1	1.5	2.25	0.5	1.25	2.5
1	1	1.2	1.44	0.2	0.44	2.2
1	1	1.1	1.21	0.1	0.21	2.1
1	1	1.05	1.1025	0.05	0.1025	2.05
1	1	1.01	1.0201	0.01	0.0201	2.01

c) The gradient at the point $(1,1)$ is 2.

d) The gradient at the point $(-1,1)$ is -2.

Progress questions (3)

1. (a)

A		B				Gradient of AB
x	y	x	y	dx	dy	dy/dx
2	4	3	9	1	5	5
2	4	2.1	4.41	0.1	0.41	4.1
2	4	2.05	4.2025	0.05	0.2025	4.05
2	4	2.01	4.0401	0.01	0.0401	4.01

The gradient at $(2,4)$ is 4

(b)

A		B				Gradient of AB
x	y	x	y	dx	dy	dy/dx
−3	9	−4	16	−1	7	−7
−3	9	−3.1	9.61	−0.1	0.61	−6.1
−3	9	−3.05	9.3025	−0.05	0.3025	−6.05
−3	9	−3.01	9.0601	−0.01	0.0601	−6.01

The gradient at $(-3,9)$ is -6.

2.

x	gradient (dy/dx)
1	2
2	4
−3	−6

3. $dy/dx = 2x$

Practical assignment (3)

The point of contact is $(1.5, 2.25)$ and can be found by solving the equation $x^2 = 3x - 2.25$
The gradient of both curve and line at the point of contact is 3.
This fits with our previous deduction, $dy/dx = 2x$.

Practice questions (4.1)

(a)

A		B				Gradient of AB
x	y	x	y	dx	dy	dy/dx
1	1	1.2	1.728	0.2	0.728	3.64
1	1	1.1	1.331	0.1	0.331	3.31
1	1	1.05	1.157625	0.05	0.157625	3.1525
1	1	1.01	1.030301	0.01	0.030301	3.0301

The gradient at A is 3.

(b)

A		B				Gradient of AB
x	y	x	y	dx	dy	dy/dx
2	8	2.2	10.648	0.2	2.648	13.24
2	8	2.1	9.261	0.1	1.261	12.61
2	8	2.05	8.615	0.05	0.615	12.3
2	8	2.01	8.121	0.01	0.121	12.1

The gradient at A is 12.

(c)

A		B		dx	dy	Gradient of AB
x	y	x	y	dx	dy	dy/dx
−3	−27	−3.2	−32.768	−0.2	−5.768	28.84
−3	−27	−3.1	−29.791	−0.1	−2.791	27.91
−3	−27	−3.05	−28.373	−0.05	−1.373	27.46
−3	−27	−3.01	−27.271	−0.01	−0.271	27.1

The gradient at A is 27.

Practice questions (4.2)

1. When $y = 2x^2$, $dy/dx = 4x$

   We can deduce this from the table of values in Practice Questions (2), by multiplying all the y values by 2, and completing the rest of the table again.

2. a) When $y = 2x$, $dy/dx = 2$

   Since $y = 2x$ is a straight line, we can use the method shown at the beginning of chapter 2 to find the gradient, which is the same at every point on the line.

   b) $dy/dx = −1$

   c) $dy/dx = k$

3. a) When $y = 2$, $dy/dx = 0$. The line $y = 2$ is horizontal, so its gradient is zero at every point.

   b) $dy/dx = 0$

   c) $dy/dx = 0$

Progress question (4)

y	dy/dx
x^3	$3x^2$
ax^2	$2ax$
bx	b
c	0

Practical assignment (4)

Your curve should begin at a point on the y-axis (where x, representing time, is 0).

The gradient should be negative everywhere, since the speed of the ball will be decreasing because of the friction with the floor. (The speed cannot increase as the floor is flat).

The journey of the ball finishes when the speed is zero, so the curve stops when it reaches the x-axis.

Practice questions (5.1)

(a) $dy/dx = 2x$ (b) $dy/dx = 3$ (c) $dy/dx = a$

(d) $dy/dx = 0$ (e) $dy/dx = 0$ (f) $dy/dx = 8x$

(g) $dy/dx = 6x^2$ (h) $dy/dx = 0$ (i) $dy/dx = 4x^3$

Practice questions (5.2)

(a) $dy/dx = 2x - 8$ (b) $dy/dx = 8x + 4$

(c) $dy/dx = 3x^2 + 12x + 12$ (d) $dy/dx = 2$

(e) $dy/dx = -2/x^3$ (f) $dy/dx = x^{-\frac{1}{2}}$

(g) $dy/dx = \frac{1}{4} x^{-\frac{3}{4}}$ (h) $dy/dx = -\frac{1}{2}x^{-3/2}$

(i) $dy/dx = 1 - 3/x^2$ (j) $dy/dx = \frac{1}{2}$

(k) $dy/dx = -4x^{-5}$ (l) $dy/dx = \frac{3}{4} x^{-\frac{1}{4}}$

(m) $dy/dx = 1 - x^{-\frac{1}{2}}$ (n) $dy/dx = \frac{1}{2}x^{-3/2}$

(o) $dy/dx = \frac{1}{2}x^{-\frac{1}{2}}$ (remember that $\sqrt{x} = x^{\frac{1}{2}}$)

Progress questions (5)

1. (a) (0.5,2.75)
   (b) (1,4)
   (c) (0.707,0.354), (−0.707,−0.354)
2. (a) $dy/dx = 3(x + 2)^2$
   (b) $dy/dx = 9(3x - 1)^2$

No. If you take the same short cut in (b) a factor of 3 is missing.

3. (a) Yes
   (b) No
   (c) Yes

(More on this in Chapter 10).

Practical assignment (5)

(a) First, notice from your sketch that the gradient of $y = 1/x$ is negative at every point on the curve.
(b) (i) When $x > 0$, the gradient of $y = 1/x$ increases as x increases.
   (ii) When $x < 0$, the gradient of $y = 1/x$ increases as x decreases.
      Note that in both the first and the third quadrants, the curve approaches the x-axis and the gradient approaches zero.
(c) Your sketch should look similar in shape to the curve $y = -1/x^2$ (see below).

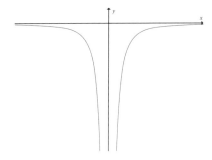

Fig A5.1

(d)
(e) See (c)!

Practice questions (6.1)

(a) $dy/dx = 6x + 5$ $d^2y/dx^2 = 6$
(b) $dy/dx = 3x^2/2$ $d^2y/dx^2 = 3x$

(c) $dy/dx = 8x^3 - 3x^2$ $d^2y/dx^2 = 24x^2 - 6x$
(d) $dy/dx = -2/x^2$ $d^2y/dx^2 = 4/x^3$
(e) $dy/dx = -2/x^3$ $d^2y/dx^2 = 6/x^4$
(f) $dy/dx = 1$ $d^2y/dx^2 = 0$

Progress questions (6)

1. (a) (0,0)   (b) 0   (c) No   (d) 0
2. (a) (0,2)   (b) 2   (c) No   (d) 0

Practical assignment (6)

1. (a) (i) The gradient is small and negative
   (ii) The gradient is small and positive
   (b) The gradient of the gradient at the origin is positive (value 2), (and has the same value everywhere)
2. (a) (i) The gradient is small and positive
   (ii) The gradient is small and negative
   (b) The gradient of the gradient at the origin is negative (value –2), (and has the same value everywhere)
3. (a) (i) The gradient is small and positive
   (ii) The gradient is small and positive.
   (b) The gradient of the gradient at the gradient at the origin is zero, but does not have the same value everywhere. For $x < 0$, the gradient of the gradient is negative. For $x > 0$, the gradient of the gradient is positive. The numerical value (also called the "absolute" value) of the gradient of the gradient increases as x gets further away from zero.

Practice questions (7.1)

1. $y = x^3$ has a point of inflexion at (0,0). The gradient is positive on each side of this point.
2. $y = -x^3$ has a point of inflexion at (0,0). The gradient is negative on each side of this point.
3. $y = 1/x$ has no points with zero gradient.

4. $y = x^3 - 3x$ has a maximum at $(-1,2)$ and a minimum at $(1,-2)$.

Progress questions (7)

1. (i) The gradient function of $y = x^3$ changes from positive to zero to positive again around the point of inflexion $(0,0)$. This means that the gradient function itself has a minimum. The second derivative d^2y/dx^2 will therefore be zero at $(0,0)$. We can verify this:

$y = x^3$
$dy/dx = 3x^2$ (gradient always positive)
$d^2y/dx^2 = 6x$ so when $x = 0$, $d^2y/dx^2 = 0$

(ii) As (i), but for positive read negative.

$y = -x^3$
$dy/dx = -3x^2$ (gradient always negative)
$d^2y/dx^2 = -6x$ so when $x = 0$, $d^2y/dx^2 = 0$

2. $dy/dx = 3x^2 + 6x - 9$

For stationary points, $dy/dx = 0$:

$$3x^2 + 6x - 9 = 0$$

Dividing through by 3:

$$x^2 + 2x - 3 = 0$$
$$(x + 3)(x - 1) = 0$$
$$x = -3, x = 1$$

Stationary points are $(-3,26)$ and $(1,-6)$

$d^2y/dx^2 = 6x + 6$
When $x = -3$, $d^2y/dx^2 = -12$
This is a negative value, so $(-3,26)$ is a maximum
When $x = 1$, $d^2y/dx^2 = 12$,
This is a positive value, so $(1,-6)$ is a minimum.

Practical assignment (7)

This table shows the results for cut-out corners of lengths 1 to 6:

corner	length	width	volume
1	18	10	180
2	16	8	256
3	14	6	252
4	12	4	192
5	10	2	100
6	8	0	0

The next thing to do would be to "zoom in" on values around 2, and we get these results:

corner	length	width	volume
1.8	16.4	8.4	247.97
1.9	16.2	8.2	252.40
2	16	8	256.00
2.1	15.8	7.8	258.80
2.2	15.6	7.6	260.83
2.3	15.4	7.4	262.11
2.4	15.2	7.2	262.66
2.5	15	7	262.50

We have now done quite a lot of calculations but we have not found the "perfect" solution, i.e. the size of the cut-off corner which will give the biggest possible volume. We can see it will be near 2.4, and we could continue using this trial and improvement method for values in this range. However, calculus provides us with a much more sophisticated approach – no time will be spent on irrelevant values, and we can get straight to the correct solution.

Let the size of the cut-off corners be x cm by x cm.

So the length of the box is $(20 - 2x)$ and the width is $(12 - 2x)$.

The volume is length × width × height, so the expression for the volume is:

$$V = x(20 - 2x)(12 - 2x)$$

To find the value of x which gives the maximum volume, we need to differentiate this expression. First, we multiply out the brackets and simplify, giving:

$$V = 240x - 64x^2 + 4x^3$$

Differentiation gives:

$$dV/dx = 240 - 128x + 12x^2$$

For a maximum (or minimum) value, we need dV/dx to be zero, which gives this equation:

$$12x^2 - 128x + 240 = 0$$

which simplifies to:

$$3x^2 - 32x + 40 = 0$$

This equation cannot be factorised. This is what we expect, because we know the solution we are looking for is not a whole number or a simple fraction. Applying the quadratic equation formula gives:

$$x = \frac{32 \pm \sqrt{304}}{6}$$

and the two values of x are 2.427 and 8.239 (correct to 3 decimal places).

Clearly, 2.427 is the required solution, because our calculations pointed to a value close to 2.4. This is a practical application, and we discard the other solution.

But it is important to understand why two solutions were obtained, even though we are only interested in one of them. The expression for the volume is a cubic function. You have seen from curve sketching that in general, a cubic function has two stationary points, one maximum and one minimum. (The curve $y = x^3$ is a special case, since its maximum and minimum coincide in a point of inflexion). In our current problem, $x = 8.239$ would give the minimum point on the curve, and we ignore this because we are not interested in minimising the volume. In any case, the **practical** minimum volume is zero, which occurs when $x = 6$ (look back at the table of values).

If we had gone directly to the calculus method, with no preliminary calculation, how would we have known which solution gives the maximum volume? We would use the rule which tells us that at a maximum point, the second derivative will be negative. We would find the second derivative:

$$d^2V/dx^2 = 6x - 32$$

and then calculate the value of this function for each of the two solutions:

x	$6x - 32$
2.427	−17.438
8.239	17.434

Since the second derivative is negative when $x = 2.427$, this is the required solution. We can find the volume of the box by putting $x = 2.427$ in the expression for V:

$$V = 2.427 \times (20 - 4.854) \times (12 - 4.854) = 275.916$$

In practice, we can't make this box because we can't measure cardboard to 3 decimal places of centimetres. Nevertheless, the calculus method takes us straight to a solution and we can then choose an appropriate degree of accuracy.

Practice questions (8.1)

1. Using the two walls as two sides of a rectangle measuring 30 by 70 is not the best solution. We would have to use our 100 metres of fencing to form the other two sides,

and the rectangle would have an area of 2100 square metres. We saw in the first example that a longer, thinner, rectangle, gives a smaller area.

So, we must investigate the possibility of using some of our 100 metres of fencing to extend the 30-metre wall, but in this case we will not have enough fencing to make use of all of the 100 metre wall.

(see Fig A8.1)

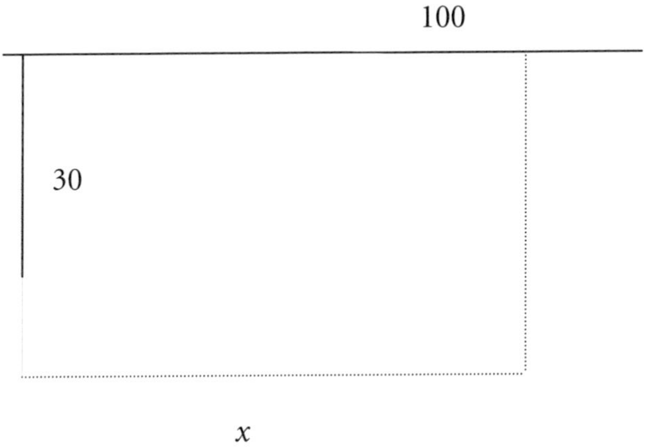

100

30

x

Fig A8.1

Let the length of the rectangle be x.
Let the extension to the 30 metre wall be e.
The lengths of the 3 sections of fencing are e, x, and $(30 + e)$.
  So $e + x + 30 + e = 100$

$$2e + x = 70$$

$$e = \frac{(70 - x)}{2}$$

Area, $A = x(30 + e) = x\left(30 + \frac{(70-x)}{2}\right)$
This expression simplifies to $A = 65x - \frac{x^2}{2}$
Then $dA/dx = 65 - x$, and when $dA/dx = 0$, $x = 65$, $e = 2.5$
and $A = 65 \times 32.5 = 2112.5$ sq. metres.

This is not a dramatic improvement on a 70 by 30 enclosure with area 2100 square metres. But the point is that calculus can provide the **best** solution, not the nearly best!

2. Let x be the length of the side of the cut-off squares.
Then the expression for the volume is:

$$V = x(10 - 2x)^2$$

Expanding the bracket gives:

$$V = 100x - 40x^2 + 4x^3$$

Differentiating, $dV/dx = 100 - 80x + 12x^2$
Putting $dV/dx = 0$ gives us a quadratic equation with
solutions $x = 5$ and $x = 5/3$.

V must be a maximum when $x = 5/3$. The other solution,
$x = 5$, certainly gives the minimum volume, as in this
case the box would have zero dimensions! Mathematical
confirmation of which values give a maximum or minimum
would be obtained by looking at the value of d^2V/dx^2:

$$d^2V/dx^2 = -80 + 24x$$

x	$-80 + 24x$		
5	40	positive	minimum
5/3	-40	negative	maximum

Progress questions (8)

a) $r = 3$ $h = 250/9\pi = 8.84$ (to 2 d.p.)
   $A = 2\pi \times 9 + 500/3 = 223.21$ (to 2 d.p.)
b) $r = 3.4$ $h = 6.88$
   $A = 219.69$
c) Volume of aluminium $= A \times 0.5$

r	V
3	111.61
3.4	109.85

Saving on aluminium $= 1.31$ cm^3
% saving $= 1.31/111.61 \times 100 = 1.17$ %

This might seem rather a lot of trouble for a small result.
   But aluminium would be a large part of the drinks company's
production costs – thousands of dollars per year – so 1.17% of
these costs would be a significant amount of money.

Practical assignment (8)

Let w be the width of the tank, and h the height.
Then the area of metal is given by:

$$A = 4wh + w^2$$

The volume, V, is 4000 litres, which gives:

$$w^2h = 4000$$

We need to eliminate one variable so that A becomes a function of one variable only. So write h in terms of w:

$$h = 4000/w^2$$

and substitute this value into the expression for A:

$$A = (16000/w) + w^2$$

Differentiate:

$$dA/dw = (-16000/w^2) + 2w$$

Equating this expression to zero leads to:

$$w = 20, h = 10 \text{ and } A = 1200.$$

Practice questions (9.1)

1. $y + 4x + 4 = 0$.
2. $4y = 3x - 1$
3. The line $y = mx + c$ and the curve $y = x^2 + c$ both intersect the y-axis at the point $(0,c)$.
   For the line to be a tangent to the curve, it must touch the curve at one point, and this must be the point $(0,c)$.
   Hence the tangent is parallel to the x-axis, and $m = 0$.

Practice questions (9.2)

1. $y = -2x + 7$
2. (i) $2y + x = 3$
   (ii) $4y = x + 18$
3. $24y + 32x = 19$
4. (i) $(1,1)$ and $(-1,-1)$
   (ii) $x + y = 2, x + y = -2$.
5. (i) $(-1,0), (0,0), (1,0)$

(ii) $dy/dx = 3x^2 - 1$, gradient at the origin is -1

(iii) Since the gradient at the origin is negative, the curve is above the x-axis between $x = -1$ and below the x-axis between $x = 0$ and $x = 1$.

(iv) The gradient of the normal at the origin is 1.
The equation of the normal is $y = x$.

(v) The normal meets the curve at the points where

$$x(x - 1)(x + 1) = x$$

One solution to this equation is $x = 0$.
(We knew that already!)
If $x \neq 0$, we can divide both sides of the equation by x, so the other solutions are given by

$$(x - 1)(x + 1) = 1$$
$$x^2 - 1 = 1$$
$$x^2 = 2$$
$$x = \pm\sqrt{2}$$

Substituting $+\sqrt{2}$ and $-\sqrt{2}$ into the function gives the corresponding y-values $+\sqrt{2}$ and $-\sqrt{2}$, so the points are P $= (\sqrt{2},\sqrt{2})$ and R $= (-\sqrt{2},-\sqrt{2})$.

(vi) The gradient at P is the value of dy/dx when $x = \sqrt{2}$.
$dy/dx = 3x^2 - 1$
At P, $dy/dx = 3(\sqrt{2})^2 - 1 = 5$
By symmetry, the gradient at R is -5.

Progress questions (9)

1. (i) M $= (0,-1)$ N $= (0,3/2)$ (ii) MN $= 2\frac{1}{2}$
2. (i) P $= (\frac{1}{2},3\frac{1}{2})$ (ii) a $= 3\frac{1}{4}$ (iii) $x + y = 4$
3. (i) N $= (-1/6,\frac{1}{2})$ (ii) $k = 19/36$ (iii) $y = \frac{1}{3}x + 5/9$

Practical assignment (9)

$$y = x^2$$
$$y = -\frac{1}{2}x + 3/2$$
$$x^2 + \frac{1}{2}x - 3/2 = 0$$
$$x = 1 \text{ or } x = -3/2$$

The point of intersection in the 2nd quadrant is $(-3/2, 9/4)$

Practice questions (10.1)

(a) $dy/dx = 6(x-2)^5$ (b) $dy/dx = 4x(x^2+1)$

(c) $dy/dx = \frac{-(2x+2)}{(x^2+2x)^2}$ (d) $dy/dx = 4(2x-3)(x^2-3x)^3$

(e) $dy/dx = \frac{-3(1-x^{1/2})^2}{x^{1/2}}$ (f) $dy/dx = 2(x+1/x)(1-1/x^2)$

(g) $dy/dx = (1+2x)^{-1/2}$ (h) $dy/dx = -3(3x^2-1)(x^3-x)^{-4}$

(i) $dy/dx = 3x(\tfrac{1}{2}x^2+1)^2$ (j) $dy/dx = 8(2x-3)^3$

Practice questions (10.2)

(a) $dy/dx = \frac{4x-1}{2y}$ (b) $dy/dx = 2/y$
(c) $dy/dx = 1/y^2$ (d) $dy/dx = -x/y$
(e) $dy/dx = 1/3y^2$ (f) $dy/dx = 1/4x^{1/2}y$
(g) $dy/dx = 1/4y^2$ (h) $dy/dx = -3x^2/y^2$

Progress questions (10)

1. (a) $dy/dx = 2(2x+2)(x^2+2x+1)$
   (b) $dy/dx = 4(x+1)^3$
   (c) $y = (x+1)^4$; $dy/dx = 4(x+1)^3$
2. (a) $dy/dx = 2x(x+2)$
   (b) $y = (x^2+2x)^2$; $dy/dx = 2(x^2+2x) = 2x(x+2)$

Practice questions (11.1)

1. (a) $dy/dx = 6x$ (b) $dy/dx = 6x$ (c) $dy/dx = 6x$
2. $y = 3x^2 + c$ (c is any positive or negative number or zero)
3. (a) $dy/dx = 4x$ (b) $dy/dx = 4x$ (c) $dy/dx = 4x$
4. $y = 2x^2 + c$ (c is any positive or negative number or zero)

Practice questions (11.2)

(a) $y = 6x^2 + c$ (b) $y = x^2 - 3x + c$ (c) $y = 2x^2 + 2x + c$

Practice questions (11.3)

(a) $y = x^3 + c$ (b) $y = \frac{x^3}{3} + c$ (c) $y = x^4 + c$

(d) $y = \frac{x^4}{4} + c$ (e) $y = x^{-1} + c$ (f) $y = -2x^{-1} + c$

(g) $y = x^n + c$ (h) $y = \frac{x^{n+1}}{(n+1)} + c$

Progress questions (11)

(a) $y = \frac{2x^9}{3} + c$ (b) $y = \frac{x^3}{3} - x + c$ (c) $y = 4x + c$

(d) $y = -\frac{1}{2}x^2 - 3x + c$ (e) $y = \frac{-1}{x} - x + c$

(f) $y = \frac{x^3}{3} + 2x^2 + 4x$ (g) $y = \frac{x^{2n+1}}{2n+1}$ (h) $y = \frac{2x^{(3/2)}}{2}$

Practical assignment (11)

(a) The rectangle has height a and width x. The area is ax.

(b) The triangle has base x and height x. The area is $\frac{1}{2}x^2$.

(c) In each case, the area between the given line and the x-axis can be found by integrating the function represented by the line:

Line	Area
$y = a$	ax
$y = x$	$\frac{1}{2}x^2$

Practice questions (12)

1. (i) 9   (ii) 12.5   (iii) 27
2. (i) 2.5   (ii) 2.33

Progress questions (12)

1. 3.77
2. (a) 5.33   (b) 2.67
3. 0.5

Practice questions (13)

1.

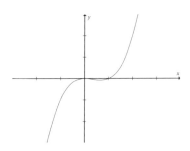

Fig A13.1

(i) The gradient is negative when $0 < x < 2/3$

(ii) The gradient is zero at the origin and at the point where $x = 2/3$

2.

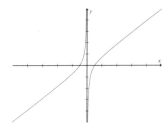

Fig A13.2

(i) $dy/dx = 1 + 1/x^2$; since x^2 is always positive, dy/dx is always positive.

(ii) As x gets larger, $1/x^2$ gets smaller.
The limiting value of $1/x^2$ is 0, so the limiting value of the gradient is 1.

3. (i) The minimum value of $(x-2)^2$ is 0, since a squared value cannot be negative.
So the minimum value of the function is 3.

(ii) The minimum value occurs when $x = 2$.

(iii) 7

(iv)

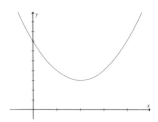

Fig A13.3

(v) 4

(vi) 10.67 (to 2 d.p.)

(Find the area between the curve and the x-axis and subtract from the area of the surrounding rectangle).

4. (i) $y = 2x^2 - 3x + 1$ (ii) $x = \frac{1}{2}$ (iii) $(3/4, -1/8)$
    (iv)

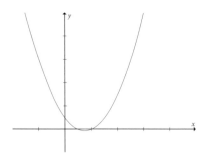

Fig A13.4

(v) $-1/24$ or -0.417 to 3 d.p.

The area is negative because the curve is below the axis; the y-values are negative in this range of x.

5. (i) 24   (ii) 12   (iii) 49
    (iv) at time $t = 5$ seconds

(v)

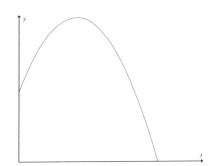

Fig A13.5

6. (i) $dy/dt = 3t^2$ $dt/dx = 2x$ (ii) $6x (x^2 - 1)^2$
   (iii) $y = (x^2 - 1)^3$ (iv) $dy/dx = 6x (x^2 - 1)^2$
   (v) The chain rule.

7. (i) $(x - 1)(x + 1) (x - 2)$; $x = 1, x = -1, x = 2$
   (ii) 2
   (iii)

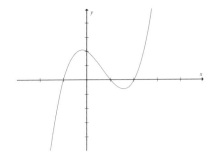

Fig A13.6

   (iv) $x = 1, x = 1/3$ (v) $y = 3x - 6$; $3y = x + 2$
   (vi) 6.67 (to 2 d.p.)

8. (i) 13/12 (1.083 to 3 d.p.) (ii) −5/12 (−0.417 to 3 d.p.)
   The area calculation in (ii) gives a negative result because
   the curve is below the x-axis. The absolute (positive) value
   of this area is 0.417. Integration using the limits 0 and 2
   would give 13/12 − 5/12 = 2/3 which is not correct for the
   total area. The correct result is 13/12 + 5/12 = 1½ .

The only way to deal with this type of problem is first to determine where the curve cuts the x-axis, secondly to sketch the curve, integrate each section separately, and finally add the positive values of the areas.

9.   (i) (0,5) and (4, –28).   (ii) max and min respectively.
     (iii) $x = 1$ (iv) $(x-1)(x^2-5) = 0$; $1, \sqrt{5}, -\sqrt{5}$
     (v)

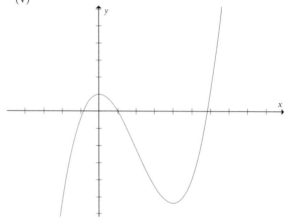

Fig A13.7

10. (i) (0,0) (½, –1/8) (–½, –1/8)
     (ii)

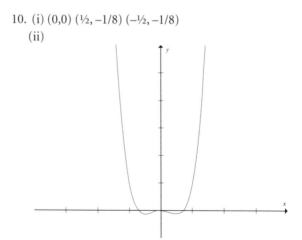

Fig A13.8

(iii) "The number of stationary points on the curve of an algebraic function is one less than the highest power of x in the function, unless a maximum and a minimum point coincide at a point of inflexion, in which case the number of stationary points is reduced by one".

11. $a = 10$ cm, $m = 5.774$ cm (to 3 d.p.)

12. (i)

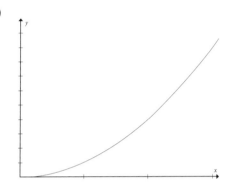

Fig A13.9

    (ii) 16/3 (5.33 to 2 d.p.)   (iii) $h = 9$.

13. (a) (i) $dy/dx = -x/y$ (ii) $dy/dx = -x/\sqrt{(r^2 - x^2)}$
     (b) Using (a)   (ii) $dy/dx = -0.5$ gives:

$$x/\sqrt{(r^2 - x^2)} = 0.5$$

Rearranging this equation and squaring both sides gives:

$$(2x)^2 = r^2 - x^2$$

and the solutions of this equation are $x = r/\sqrt{5}$ and $x = -r/\sqrt{5}$

Putting $r = 5$, we have $x = \sqrt{5}$ and $x = -\sqrt{5}$

The corresponding y–coordinates are $2\sqrt{5}$ and $-2\sqrt{5}$

so the points on the circle are $(\sqrt{5}, 2\sqrt{5})$ and $(-\sqrt{5}, -2\sqrt{5})$.

Of course you should have checked that these co-ordinates satisfy the equation of the circle, $x^2 + y^2 = 5$.

14. (ii) The first three (x,y) pairs give these equations:

$$\frac{a}{4} + \frac{b}{2} + c = 0 \; ; \quad 4a + 2b + c = 3; \quad 9a + 3b + c = 10$$

Solving these equations simultaneously gives $a = 2$, $b = -3$, and $c = 1$.

So the supposed quadratic function is $y = 2x^2 - 3x + 1$.

(iii) Putting $x = 3.5$ gives $y = 15$, and putting $x = 6$ gives $y = 55$.

The quadratic relationship is confirmed for all the observed values.

(iv) The minimum value of y is $-\frac{1}{8}$, which is found by differentiation in the normal way.

15. (a) (i) $dy/dx = \frac{2x-2}{y}$

(ii) The co-ordinates of the stationary point are $(1,3)$

(iii) $x = 2$, $x = 0$.

(iv)

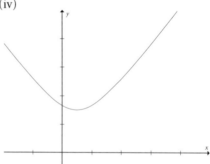

Fig A13.10

(b) (i) $k = 6$

(ii) As k decreases, the curve moves vertically downwards and its minimum point approaches the x-axis.

(iii) If $k = 2$, the function becomes $y^2 = 2x^2 - 4x + 2 = 2(x^2 - 2x + 1) = 2(x-1)^2$.

Then we can take the square root of each side and the function becomes $y = \pm\sqrt{2}(x-1)$.

But remember we are only considering positive values of y.

This means that when $x < 1$, we take $y = -\sqrt{2}(x-1)$, giving points $(0.5, 0.707)$, $(0, 1.414)$ etc.

The graph of the function now becomes a pair of straight lines (see FigA13.11).

The point (1,0) is not a minimum because the gradient is not zero at this point, in fact the gradient is not defined at the point (1,0). We say that the function *is not differentiable* at (1,0).

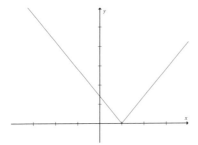

Fig A13.11

16.

(i)

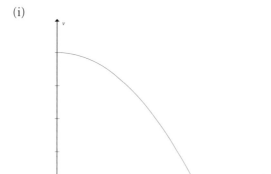

Fig A13.12

(ii) $v = 3$ when $t = 1$

(iii) 8 metres

(iv) 4 metres

(v) If the deceleration (dv/dt) were constant (see FigA13.13) then the function would be $v = 4 - 2t$ and the average speed would be 2 m/s.

The value of the function $v = 4 - t^2$ is greater than the value of the function $v = 4 - 2t$ at every point in the range $0 < t < 2$, so the average speed is greater than 2.

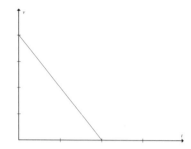

Fig A13.13

   (vi) The distance must be less than 8 metres (see part (iii)).
At an average speed of 2 m/s , the distance would be 4 metres.

      So the distance is between 4 and 8 metres.

  (vii) Integration of the function $v = 4 - t^2$ in the range $0 < t < 2$ gives distance = $5\frac{1}{3}$ metres.

17. (i) $t = 5$.

   (ii) 333 metres

  (iii) We want $v = 0$ and $t = 3$. These values give $k = 11.1$ (recurring).

      If $k = 11$, $t = 3.015$ (too big). If $k = 12$, $t = 2.887$.

      So the required whole number value of k is 12.

  (iv) 192 metres.

18. (i) $(2,0)$ $(5,0)$

   (ii) $(0,10)$

  (iii) $(3.5, -2.25)$ This is a minimum.

  (iv)

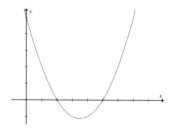

Fig A13.14

(v) $8\frac{2}{3}$

(vi) $-4\frac{1}{2}$

(vii) The answer is negative because the area is below the x–axis.

19. (i) $n = 100 - 8p$

(ii) $P = n(p - 2)$

(iii) $P = 116p - 8p^2 - 200$

(iv) $p = £7.25$, $n = 42$, $P = £220.50$

(v) As £7.25 is a slightly awkward amount of money, you might advise the window cleaner to charge £ 7.00 or £7.50. Either of these prices, in theory, will yield a profit of £220.

20. (i) $dy/dx = 3x^2 - 6$.

(ii) 6.

(iii) Near to, and each side of the point (0,4), the gradient is just less than 6.

(iv) $d^2y/dx^2 = 6x$. At the point (0,4), $d^2y/dx^2 = 0$.

(v) At (0,4) the gradient reaches a maximum value of 6. This is confirmed by the fact that $d^2y/dx^2 = 0$ at this point.

(vi)

Fig A13.15

135

Glossary

Area The amount of space, measured in square units, inside a two-dimensional shape.

Calculus A set of mathematical tools which enables us to investigate changing quantities and make accurate mathematical statements about these quantities.

Chord A line joining two points on a curve.

Coefficient The constant multiplier of a term in an algebraic function.

[Example: in the function $y = x^2 - 2x$, 3 is the coefficient of x^2 and 2 is the coefficient of x.]

Constant A number which is fixed, but may be known or unknown.

Cubic A cubic function contains terms in powers of x, of which the highest is 3.

Derivative A function which is obtained from another function by differentiation.

Differentiation The mathematical process which enables us to find the gradient of a curve at a chosen point on the curve.

Estimate An approximate value deduced by observation.

Gradient The "steepness" of a straight line, or of a tangent to a curve. It is calculated by treating the sloping line as the hypotenuse of a right-angled triangle, and taking the tangent of the angle of elevation (which means dividing the vertical height of the triangle by its width).

Hyperbola A curve which represents the function $y = 1/x$, or a related function (e.g. $y = x + 1/x$). The curve has two disconnected sections which approach, but never touch, two straight lines. In the case of the rectangular hyperbola, these two straight lines are the x- and y-axes.

Integration The mathematical process which enables us to find the function whose gradient function is given (though this is not possible in all cases). Integration is also used to find the area of a shape with one or more curved edges.

Limit A value which may be **approached** by calculation but is usually deduced by reasoning.

Normal A straight line at right angles to a tangent, cutting the curve at the point where the tangent touches it.

Notation The general term for the symbols used to represent mathematical equations, values, etc. Notation is established by common practice.

Perimeter The distance around the boundary of a two-dimensional shape.

Quadrant The x,y plane in which we plot graphs is divided into four sections by the x- and y-axes. The first quadrant is where x and y are both positive. The remaining quadrants are numbered by moving anti-clockwise round the origin.

Quadratic A quadratic function contains terms in powers of x, of which the highest power is 2.

Stationary point A point on a curve where the gradient is zero.

Tangent A straight line which **touches** a curve at one point. It may cut the curve elsewhere, but normally we are interested in the point where it touches the curve.

Websites

Here are some sites which may help you with the topics covered in this book:

www.math.about.com/cs/calculus/index.htm

Excellent site with detailed tutorials on a range of maths topics. The calculus module breaks down into 8 parts including a test. Everything is explained using easy-to-understand examples. Animated diagrams.

www.mathnerds.com

A site which invites you to send in a question on any maths topic. It's free but you have to submit a few details about yourself. They say they normally give a quick response – immediate, or maximum two days.

www.bbc.co.uk/scotland/education/higher/maths/calculus/index.shtml

Colourful, fun site. Animated and interactive. Gives a series of activities for each topic. Aims at achieving understanding rather than practice.

www.gazinotes.com/a-level/a_level_mathematics.htm

A survival guide – general advice about how to study maths.

www.mathsnet.net/asa2/2004/index.html

Excellent site with lots of clear explanation, animated graphs, and exam questions with solutions shown in stages. The maths content is preceded by a glossary and an explanation of symbols.

www.s-cool.co.uk/topic_index.asp?subject_id=1&d=0

Very good site with good explanations of a wide range of maths topics.

The following sites will be helpful for revision of more elementary maths:

www.bbc.co.uk/schools/gcsebitesize/maths

www.easymaths.com/problems_main.htm

Index